GOVERNMENT PROMOTION OF AMERICAN CANALS AND RAILROADS

1800-1890

CARTER GOODRICH

GOVERNMENT PROMOTION OF AMERICAN CANALS AND RAILROADS 1800-1890

COLUMBIA UNIVERSITY PRESS
NEW YORK 1960

PUBLISHED IN GREAT BRITAIN, INDIA, AND PAKISTAN
BY THE OXFORD UNIVERSITY PRESS
LONDON, BOMBAY, AND KARACHI

Library of Congress Catalog Card Number: 60-6546
MANUFACTURED IN THE UNITED STATES OF AMERICA

FOR

FLORENCE NIELSEN GOODRICH

PREFACE

Publication of a book on government promotion is made timely by the current interest in economic development. During the century in which the United States transformed itself into a great industrial nation, the most important and the most widely employed measures by which American governments attempted directly to promote economic growth were devoted to the construction of canals and railroads. The analysis of this experience may be of service to those who are considering what should be the roles of government and of private enterprise in the nations now striving for development.

The writing of the book has been made possible by the large amount of monographic research that has been carried on in recent years by a number of scholars, particularly on the activities of state and local governments. A decade or two ago, a general book on the subject could hardly have been ventured. Now, although much remains to be done in the collection and analysis of data, it appears possible to present the main outlines of the history of public promotion in nineteenth century America and to offer some interpretation of the part it played in the economic development of the country.

The study has received generous support from the Council for Research in the Social Sciences of Columbia University, which made a series of grants beginning in 1946. Earlier findings have been published in four articles in the *Political Science Quarterly*, three in the *Journal of Economic History*, and two in the *Pro-*

ceedings of the American Philosophical Society. The titles are listed in the Bibliography. With the kind permission of the editors, the substance of these articles has been freely used in the present work. Aid in the final stages of the project has been provided by the Industrial Countries Center of the Economics Research Centers of Columbia University.

For stimulating criticism through the entire period of the study, I owe much to my friend and colleague Joseph Dorfman. I have also been encouraged by the interest taken in the work by fellow-members of the Committee on Research in Economic History and of the Economic History Association, of whom a number have worked longer than I in the field of internal improvements.

A particularly rewarding aspect of the work has been association with successive groups of graduate students at Columbia whose research has been devoted to related subjects. H. Jerome Cranmer, Harvey H. Segal, and Nathan Miller each held a fellowship from the Committee on Research in Economic History, acted as my research assistant, and completed a dissertation on internal improvements. Other students whose dissertations or master's essays have been used and cited below are Frederick K. Abbott, James P. Cairns, George A. Donely, Robert W. Fogel, Forest G. Hill, William Marcuse, Beatrice G. Reubens, Julius Rubin and Alfred G. Smith, Jr. Vivien Carlip and Naomi Waxman Perlman served as research assistants, as did Ludwig Wagner and Jerome B. Komisar for shorter periods. Mary C. Magill and Jane Lancaster supplied cartographic skill. Carolyn M. Stedman performed the laborious task of typing without losing her curiosity as to what was to come next.

Finally, I wish to express my appreciation for the courtesies shown to me by the officials of the Press and my pleasure that their decision to publish the manuscript made the entire undertaking an enterprise of Columbia University.

CARTER GOODRICH

Inlet, New York
August 4, 1959

CONTENTS

INTRODUCTION

FEDERAL DEBATE AND DECISION
1801–1830

THE EMULATION OF STATES AND CITIES
1815–1861

FEDERAL DEBATE AND DECISION
1850–1872

THE EMULATION OF STATES AND CITIES
1861–1890

CONCLUSION

INTRODUCTION

Prompted by these actual observations, I could not help taking a more contemplative and extensive view of the vast inland navigation of these United States, and could not but be struck with the immense diffusion and importance of it; and with the goodness of that Providence which has dealt his favours to us with so profuse a hand. Would to God we may have wisdom enough to improve them.

George Washington, in a letter of 1783

1

THE SPIRIT OF IMPROVEMENT

This is a study of the role of American governments—federal, state, and local—in the creation of the facilities of inland transport. In today's terms, it is an analysis of an important aspect of development policy, the provision of social overhead capital, and of the relations between public promotion and the efforts of private enterprise. In the language of the time, the issue was that of internal improvements. The older phrase carries the connotation which the modern student must not overlook—of a movement that called for the exercise of public spirit as well as the search for immediate economic gain. To improve the country's natural advantages by developments in transportation was, in the eyes of Washington and many others, a duty incumbent both on governments and on individual citizens.[1] Even before the American Revolution, the American Philosophical Society maintained a standing Committee on American Improvements, and its members went out one winter into the frozen swamps to study the prospects of a canal to connect the Chesapeake and the Delaware. They were later praised as men "whose views, extending beyond themselves, [were] employed upon objects of general benefit and utility."[2] When canal and railroad companies were organized, individuals were often urged to buy stock not only for the dividends they would receive, but also for the satisfaction of bearing an honorable part in a great state or national work. One railroad president even reproved his stockholders for clamoring for dividends "as if they had invested as capitalists." [3]

Again and again, during a century of canal and railroad building, arguments for government support were based on appeals to the spirit of improvement. The announced purpose of Virginia's plan of state partnership with local undertakings, for example, was "to elicit individual wealth for public improvement." [4] The works themselves were thought of not only as business enterprises but also as great civic undertakings and "monuments of renown" to be passed on to future generations. Something of this emphasis survived even after it became evident that government aid was sometimes the result of public corruption rather than of public spirit; and the promoters who made the quick and questionable profits of the Credit Mobilier appeared anxious to convince the Congress and the people, not merely that they were guiltless of wrongdoing, but that they deserved the nation's gratitude for the enterprise with which they had pushed through the first transcontinental railroad.

This book will not attempt to follow the spirit of improvement through all of its manifestations. The Philosophical Society's committee gave as much attention to the importation of silkworms for the encouragement of manufactures as to the Chesapeake and Delaware Canal for the improvement of transportation. The historian of railroad aid in the South has described the movement as the last of a series of great associative efforts that began with the process of colonization itself, and a critic has pointed out that it might be regarded equally well as a forerunner of the twentieth-century activities of American governments in their aids to aviation, in their renewed programs of road building, and in the federal development of atomic energy.[5]

The present study is more limited in time and scope. Its time span is from the early days of the Republic to the completion of the American railway network toward the close of the nineteenth century. It is confined to inland transport, though somewhat similar issues are raised by government subsidies to the merchant marine and by aids to ocean navigation. Even within inland transport, it will not give equal emphasis to all forms of governmental encouragement.[6] Its principal concern is with the cases in which the users of the improvement were to pay for the services rendered. As free roads replace turnpikes, as local fee-charging navigation companies disappear, and as river and harbor improvement become services rendered without charge by the federal government, the building of roads and the clearing of rivers become of less interest to the study. Its main theme is the relation between public and private activity in the creation of canals and railroads.

Limitation of the discussion to toll- and freight-charging enterprises has the advantage of concentrating attention on those aspects of development which required deliberate choice between alternative private and public means. It was obvious that business enterprise could not be expected to provide facilities from which revenue could not be collected. But if charges were to be made for the service, and profits were a possibility, there was room for the selection of either a private or a public agency or—as the American experience showed—of a remarkable variety of combinations of the two. The focus is therefore on the controversial, and still timely, issue of competition and cooperation between government and business.

Even with these limits, the field is a wide one. The number of cases of government promotion, and the volume and importance of public investment, have been greater than is commonly remembered. Local government authorities aided turnpikes, canals or railroads in several thousand cases and in almost every state. All but a few of the state governments carried on their own programs of public improvements or gave their support to transportation companies by loans or subscriptions. Action by the federal government was less frequent and less continuous, but its history contains a period of prolonged discussion of a national plan of internal improvements and another period in which millions of dollars were loaned and more than a hundred million acres were granted for the promotion of railroad development.

These programs were indeed large enough to provide in their time a leading example of "the modern tendency to extend the activity of the state into industry." Guy Stevens Callender, who pointed this out in 1902, went on to ask why this should have taken place "here in America, where of all places in the world we should least expect to find it." [7] The question is still a challenging one to all who regard the United States of the nineteenth century as the historic stronghold of laissez faire. Yet the vigor of this American resort to government action appears somewhat less surprising if it is considered in relation to practice in other parts of the world or to the economic characteristics of developmental transportation.

The most conspicuous contrast was with England, where the railroads and most of the canals were built by unassisted private enterprise. Elsewhere, as Thorold Rogers pointed out in 1880, their construction had been "rarely thus effected . . . the state having assisted in many cases, and having altogether accomplished the work in more." [8] In France an early method of assistance to railroads was a government guarantee of return on the private capital invested. In Belgium, the state built the main railway network, leaving secondary lines to private companies. This case, together with government expenditures in Holland and Austria, was cited with approval by a committee of the United States Congress.[9] Outside of Europe, the expedient of government guarantee, "private enterprise at public risk," was used extensively in British India as well as sometimes in Brazil and Argentina. Throughout Latin America, most of the railroads were built either as public undertakings or with government subsidy. The Trans-Siberian Railway was a state enterprise, under the Czars as well as under the Soviets; and the railroads of China were for the most part built either on government account or with the aid of foreign governments establishing spheres of influence.[10] Of more direct relevance is the comparison with the nations most closely resembling the United States, the other "new countries" of British origin. Of these, Australia and New Zealand built their railroads as public enterprises; and Canada employed the three methods of government subsidy, of government construction, and of the government rescue and operation of a bankrupt private company.

Why, then, was the English example so rarely followed? The *American Railroad Journal* once demanded rhetorically why the manifest advantages of private action were not equally applicable "to the Anglo-Saxon race on this as well as on the other side of the Atlantic" and in New as well as in Old England.[11] But this spokesman for the American industry found no great difficulty in distinguishing the two cases on the many occasions on which it advocated government assistance. Certainly the circumstances under which the English canals and railroads were built differed substantially from those that prevailed either in the United States or in most other nations to which these innovations were spread. In the first place there was in England a large supply of capital available for investment. In the second place, there was more experience there than elsewhere in the use of the corporate form to mobilize and manage the resources required for large undertakings.

There was also another difference arising from the nature of the works to be attempted. A railway between London and Manchester would run through settled country and connect established channels of trade. It could expect substantial traffic as soon as it was completed. On the other hand, a railroad built westward across the plains of the United States or Canada, or eastward over the Siberian steppes, or from Sydney or Melbourne into the Australian interior, would be opening up an almost empty country and could not hope for large traffic until settlement took place along its route. The words "exploitative" and "developmental" have been suggested to describe these two types of enterprise.[12] Exploitative canals or railroads were built to take advantage of an existing opportunity and established channels of trade. With them early returns could be expected and private enterprise could operate without subsidy. On the other hand, the developmental undertaking had to depend for most of its prospective revenue on the settlement and the economic activity which its own construction was intended to promote. In the nature of the case, such development could not be immediate, and there would therefore be little prospect of substantial early profits. "Developmental transportation . . . has rarely been profitable in the narrow commercial sense of giving to those who first put up the money the 'going rate of return' on investment capital." [13] The ultimate benefits might be very great, but they would certainly be largely deferred, and much of the gain was likely to be widely diffused and inappropriable by those who invested the capital. Only a part of "all that diversified good" of which the advocates of improvement so often spoke—the increases in land value and the government's tax base, the widened opportunities for individual activities, and in general the external economies provided to agriculture and industry—would ever be collected in the tolls or freights or fares that could be charged by the enterprises themselves. Developmental undertakings, therefore, could hardly be carried to success by unaided private means. They required, at least in the absence of extraordinary illusions on the part of the original investors, either public enterprise or public assistance.

Even a cursory international comparison illustrates the relevance of these differences. A shortage of domestic capital, and the fact that governments find it easier than individuals to borrow at a distance, were throughout the nineteenth century and remain today major reasons for the governmental provision of transport facilities, particularly outside of Western Europe. Corporate or other organizations for the massing of private capital were little developed in most parts of the world. In the passage already cited, Thorold Rogers somewhat complacently attributed the resort of "foreign countries" to public enterprise to their varying degrees of deficiency in "the habits of association and enterprise." Latin American experience illustrates the distinction between exploitative and developmental enterprises, since its relatively few examples of purely private undertakings include the roads that carry coffee to Santos and sugar from the cane-fields to the Cuban ports. Bolivia provides an exceptionally exact example: the two railroads that take the tin from the great mines to the coast were built, and until very recently were owned and operated by private British interests, while the other railroads are all governmental.

Each of the three points of difference is significant for the explanation of American policy. Shortage of local capital was from the beginning a principal argument for governmental action. "The population along the contemplated line of the road," read one typical plea, "though industrious and frugal, are too poor to build the road." [14] Reviewing the history of state action for internal improvements before the United States Senate in 1850, William H. Seward declared that "a great and extensive country like this has need of roads and canals earlier than there is an accumulation of private capital within the state to construct them." [15] During the decade that followed American railroad companies were for the first time able to raise substantial sums in the European market, but lack of capital *within* a state continued to be used as an argument for government aid borrowing. "No new people," said Henry Varnum Poor, "can afford to construct their own railroads"; and the *Railroad Journal* under his authorship frequently made the distinction between the older regions, where abundance of capital made aid unnecessary, and the newer areas, where government assistance was still required.[16]

The lack of large-scale corporate enterprise was also a factor in American decisions. In the early days of internal improvements, private corporations in the United States were few in number, required special legislative acts for their incorporation, and were often attacked as instruments of monopoly and special privilege. Outside the field of banking, few of them possessed either substantial amounts of capital or experience in their management. The petty scale of much of private enterprise may be illustrated by the fact that one of the advantages claimed for the Virginia system of state aid was that its Board of Public Works, unlike the local undertakings to which it gave assistance, would have sufficient means to employ a full-time professional engineer!

Even more obvious was the developmental character of most of the projects of American improvement. There were to be sure certain particularly fortunate undertakings, principally along the Atlantic seaboard, that could exploit immediate opportunities for profitable traffic. Edward C. Kirkland compares certain of the New England railroads to wealthy dowagers who would have scorned to ask for public assistance.[17] One such dowager railroad ran from Boston to the manufacturing town of Lowell. The Delaware and Raritan, between Philadelphia and New York, might similarly be described as a dowager canal. These were exceptions, however, and through much of the nineteenth century the most characteristic American improvement was one which extended from an established center of trade and population out into a less developed area. The *Times* of London, reflecting the experience of unlucky investors, once defined the American railroad as running from "Nowhere-in-Particular to Nowhere-at-All." [18] This was only half the truth. The typical improvement was one which started somewhere and ended at a point which was perhaps nowhere when the project started, but which was to become important as a result of the improvement itself.

The force of these considerations tended to diminish as the years went on. Corporate enterprise grew in power and competence as well as in public acceptance, finding indeed in railroading a particularly favorable field of growth. The case for government action decreased in cogency as the corporations found it possible to assemble larger amounts of capital and to commit them on a longer-term basis, though sometimes the very strength of the corporations was exerted to prolong the period of subsidy. Similarly the developmental argument for support of transportation enterprises diminished in strength as population and economic activity increased. Yet developmental conditions continually recurred as settlement advanced, and the subsidy-seeking canal or railway promoter might well have appeared in Turner's famous roster of frontier types in their march across the continent. It was indeed late in this march that the necessity of crossing the Rockies and bridging some eighteen hundred miles of almost empty land between the Missouri River and the Pacific Coast became the basis for the most massive railroad assistance ever granted by the federal government.

Under these circumstances, American use of government agencies to promote the development of transportation hardly seems paradoxical, even for a nation of individualistic traditions. The spirit of improvement was abroad, and Americans were eager to build up a new country and to make their own fortunes. It need not seem strange that communities desiring rapid development should attempt to accelerate the process, even at the risk of sinking public funds in improvements that might or might not prove to be of long-run economic value. It would have been more surprising if they had been content to wait patiently for their canals and railroads until the way was clear for prudent private enterprise to supply them without assistance. The amount of governmental activity, and to some degree its distribution between periods and regions, are best to be explained as part of the great developmental effort by which the United States of the nineteenth century was opening up so vast an area of new land to settlement and economic use.

Yet neither the economics of development nor analogy from the experience of other countries would serve to account for the diversity of the expedients employed in the promotion of these American improvements. All levels of government took part in the movement, and public and civic and private efforts were brought together in an extraordinary variety of combinations.

Some commentators on internal improvements, contemporary and modern, have written as if the word governmental were a synonym for federal. A Colorado promoter, for example, boasted that he had built his railroad "without a subsidy of any kind" and then went on to recount his success in inducing three countries to issue bonds on its behalf.[19] Similarly, a railroad historian wrote that the refusal of Congress to vote assistance saved the Baltimore and Ohio from governmental interference, though the state of Maryland and the city of Baltimore were soon to hold a majority of its stock and to appoint a majority of its directors.[20] To consider internal improvements in terms of federal policy alone would miss the greater part of the story. Individual amounts of local government aid, like the subscription of Brazil Township, Clay County, Indiana, to a forgotten North and South Railroad, or even of state aid, like Virginia's assistance to the Upper Appomattox Company or to the turnpike through Snicker's Gap, were often very small. Yet the total volume of financial investment by state and local governments was considerably greater than that of the federal government, and their activities are no less instructive to the student of the relation between government and private enterprise.

At each level of government, there was a choice to be made among the major alternatives of public enterprise, of mixed enterprise in which government agencies shared ownership and control with individual investors, and of loans or donations to private corporations. The federal government built the National Road before turning it over to the states for operation and maintenance. New York State dug the Erie Canal as a public work and others followed its example. Georgia built and operated its own railroad. The city of Troy constructed a short line of railroad, and Cincinnati built one of over three hundred miles. Yet this alternative was used less often than the others. The federal government subscribed to the stock of several canal companies, and state and local authorities bought shares in turnpikes and canals and railroads. The original stockholders of the Chesapeake and Ohio Canal Company included the federal government, the state of Maryland, three municipalities, and a number of private investors.[21] Aid by loan was given by the federal government to the first transcontinental railroad, and the states and larger municipalities assisted many enterprises by loans or by guarantee of the company's bonds. There was also frequent resort to the method of outright subsidy, in amounts ranging from the small donations of villages and townships to the massive land grants made by the federal government.

When governmental investment or assistance took a directly financial form, the principal source of funds was public borrowing, though the federal government paid for its early canal subscriptions out of treasury surpluses. State enterprises like the Erie Canal and state aid to mixed or private enterprise were generally financed with borrowed funds. What the companies asked was, indeed, often described as "the loan of the state's credit." Municipal and country authorities in most cases paid their subscriptions or made their donations in bonds. But one North Carolina law, adopted at a time when the state had exhausted its borrowing capacity, authorized the counties along the line of a certain railroad to levy a tax for its benefit, payable either in cash or country produce.[22]

This was by no means the only case in which assistance was given in forms other than money or credit. The greatest largesse extended by the federal government was its gift of public lands. Texas, which entered the Union with a vast public domain, used it for the same purpose with equally lordly generosity. States and nation alike granted almost as a matter of course free right of way through public lands, often with the privilege of taking timber and stone from neighboring property. Municipalities frequently gave rights of way, as well as lots for depots or roundhouses or machine shops, and sometimes built the structures themselves. In the early days of the movement, improvement companies were sometimes permitted to raise funds by conducting lotteries, and the grant of a coveted banking franchise enabled New Jersey to secure the construction of a costly developmental canal. Exemptions from various forms of taxation were a more common and persistent form of encouragement.[23]

All this called for a complex set of legislative and administrative decisions. Government enterprises required management and operation. Assistance to private or mixed companies required the framing of conditions to ensure good faith in the completion of the work. Federal or state bonds would be issued, for example, when so many miles of track were laid; a municipal donation might be withheld until "the cars were running" to a given point. In the case of mixed enterprise, the public authorities generally, though not always, named "proxies" to vote their stock and directors to sit on the boards of the companies. Other provisions, adopted after unlucky ventures had brought losses and in some cases bankruptcy to a number of state and local governments, placed restraints on the public authorities themselves in the attempt to prevent unwise commitments. Limits were imposed on the amount of debt that could be incurred for purposes of improvement, and a vote of the electorate was generally required before local aid could be granted. In a number of cases, the development of improvement programs resulted in the creation of specialized governmental bodies. Standing committees on internal improvements were maintained in city councils in Baltimore and elsewhere, in many state legislatures, and in Congress. New York had its Canal Commissioners and Commissioners of the Canal Fund. Boards of Public Works were established to supervise the programs of a number of states, and at one time the Corps of Army Engineers was thought of as a similar agency of the national government.

The promotional activities of the many governmental agencies differed widely in the nature and scope of the public planning involved. In some cases they involved only the decision of a small community to conform or not to conform to the plans of outside promoters. In others they represented the establishment and application of a general system of aid within a state. In still others they approached more closely to the creation of a planned network of transportation facilities. In this respect public promotion ran the gamut from the mere township donation through the far-flung enterprises of some of the commercial cities and the notably coherent programs of several states to at least the serious discussion of a comprehensive national plan of internal improvements.

These diversities defy any simple analysis and can be understood only through an examination of the historical record. Even a chronological arrangement is difficult because of the variations between different areas in the same period and because of alternations of activity and retrenchment related to the ups and downs of the business cycle, the failures or successes of particular improvements, and the varying fortunes of political leaders and parties. There is, however, one preliminary distinction that can be made. State and local action, if not continuous, was at least continually recurring through the greater part of the nineteenth century, with local action persisting until its end. On the other hand, activity by the federal government on internal improvements was more sporadic and was largely confined to two periods. The first of these, an era of extensive national projects, occurred at a time when the Appalachian Mountains were the chief barrier to communication and settlement. The second, an era of extensive national subsidy, began when the railroads were being built across the prairies but reached its climax when their advance across the continent confronted the massive obstacles of the Rockies and the Sierras. It therefore seems appropriate to separate the activity of the federal government from that of state and local authorities and to begin an examination of the record with the "Era of National Projects."

FEDERAL DEBATE AND DECISION

1801-1830

. . . with me fell, I fear never to rise again, certainly never to rise again in my day, the system of internal improvement by means of national energies. . . .

With this system in ten years from this day the surface of the whole Union would have been checkered over with railroads and canals. It may still be done half a century later and with the limping gait of State legislature and private adventure.

John Quincy Adams, in a letter of 1837

2

AN ERA OF NATIONAL PROJECTS

The first decade of the nineteenth century marked the beginning of an era of national projects of internal improvement. By this time frontier settlement had passed the barrier of the Appalachian Mountains, but transportation had not yet bound the new areas to the economy of the seaboard states. In 1806 Congress authorized the first surveys for the National Road. Two years later, Albert Gallatin, the Secretary of the Treasury, presented to the Senate his Report on Roads and Canals. The first action initiated the one major developmental enterprise of the time to be constructed directly by the federal government. The second was the earliest and most distinguished attempt to formulate a comprehensive national plan of interal improvements.[1]

These actions represented a new turn in national policy. Washington's enthusiasm for improvements had been channeled through other agencies, and little attention was paid to Hamilton's statement, in the Report on Manufactures, of the advantages of federal action. The early Congresses voted appropriations for a few lighthouses and spent much time designating and redesignating the routes for post roads. Though it was sometimes recognized that "unprofitable roads might be established—from the partiality of members to their own district and country," the alternative of leaving the selection to the Executive was rejected as a greater evil and even described as a dangerous advance "toward Monarchy."[2] But it was only occasionally that voices were raised to advocate the federal construction of post roads, as against the mere choice of routes; and the most elaborate of these proposals, under which the government would have subscribed to the stock of turnpike companies for the improvement of the principal post road from Maine to Georgia, aroused little interest.[3]

On the other hand, there was a considerable background of state and local, as well as private, effort for the improvement of transportation. Throughout most of the colonial period the means of inland communication had remained extremely primitive. Roads were mainly the responsibility of local public authorities, and it was they who made the arrangements with ferrymen for the stream crossings. In the early days of scanty traffic, the towns had to give these men other inducements besides the right to collect tolls; as traffic increased, the towns were sometimes able to charge them for the privilege instead.[4] In certain cases the provincial authorities gave directions and aid for through lines of road or for roads to open up the frontier. Long afterwards, Senator Seward cited as a precedent for internal improvements an act of the early eighteenth century by which the province of New York appropriated money for a road westward from the lower Hudson, but the scale of the effort may be gauged by the requirement that the limbs of the trees had to be cut off high enough to permit the passage of a carriage with a calash top.[5] Larger projects of river improvement and canal building and of the construction of "artificial" roads were projected and undertaken in the second half of the century. "But the dark and distressing period of the revolution," as a Pennsylvania improvement society recalled, "necessarily suspended all improvements of this nature, in every part of America, until the glorious era of the peace and independence of the United States."[6]

When progress could be resumed, much of the leadership was taken by private enterprise. Before the time of Gallatin's Report two principal canals were built without financial aid from public authorities. The Santee, extending from Charleston, South Carolina, was opened in 1800, and the Middlesex, from Boston to Lowell on the Merrimack, in 1803. Each served to connect a seaport which lacked a major river with a river which lacked a satisfactory port. A number of toll-bridge companies sprang up, particularly in New England, superseding the older ferry proprietors, and were often financially successful. More important was the development of turnpike companies, of which over seventy had been chartered by 1800 and several hundred by the time of Gallatin's inquiry. These were crisscrossing the northern and middle states with a network of moderately improved toll roads, and Gallatin's informants were able to indicate certain main lines of travel formed by connecting turnpikes. These enterprises, performing "a hitherto governmental function," were operating with little or no public aid. In New England the turnpike companies received no public funds, though those in Connecticut profited by the exemption of their stock from taxation. On the other hand, New Jersey aided a turnpike company in 1804, and the city of Albany became the largest stockholder in the First Great Western Turnpike. In 1806 the Pennsylvania legislature debated the general policy of state participation and made its first subscription to the stock of a company which undertook to build a road across the mountains to the West.[7]

Meanwhile there had been a number of examples of state participation in projects of canal building and river improvement. The movement began, according to the Pennsylvania Society, "in the states of Virginia and Maryland, upon the Potomac, under the auspices of the illustrious Washington" and spread "with a noble emulation of public spirit" to other states "according to their natural advantages." Maryland and Virginia paid more than $150,000 for 340 of the 701 shares of the Potomac Company. Virginia was also interested in another and rival route to the West and in a connection with the South, and the state paid $70,000 for half of the stock of the James River Company and provided $17,500 of the $100,000 capital of the Dismal Swamp Company.[8]

The state of New York had adopted a similar policy. Two companies were organized to open the promising lines of connections leading from the Hudson River system—the Northern Inland Lock Navigation Company to develop a route to Lake Champlain and the Western to improve the Mohawk River and connect it with Lake Ontario. Each received at the outset a donation of $12,500 and later a subscription to 200 shares of stock. The Western Company, which was the more active, was also given a loan of $37,500; at the time of Gallatin's Report, the state's subscription to its stock amounted to $92,000 out of a total of $232,000. The state of Pennsylvania, which had made unusually large direct expenditures for roads and river improvements, voted a small appropriation for the Chesapeake and Delaware Canal and stood ready to pay a $300,000 bonus on the completion of the Schuylkill and Susquehanna. In the West, the state of Kentucky had committed itself to a subscription for a canal around the Falls of the Ohio.[9]

With all this emulation and stir of activity, however, the actual progress of improvement was on the whole disappointing, both in the private and the state-aided undertakings. The turnpike movement, to be sure, was still spreading, though without much financial return to the companies. The Middlesex Canal was in active use and tolls were increasing; but the Santee, though also completed, was "very unprofitable."[10] The James River Company had improved part of its river and paid a dividend. The Potomac Company had less to show. In New York the Northern Company had accomplished almost nothing; the Western Company had built and then had to rebuild a set of locks at Little Falls and had made the short connection between the Mohawk and Wood Creek. The work of both companies was virtually suspended, as was that on the Chesapeake and Delaware and a number of other undertakings. Many others, like that of Kentucky, remained still in the project stage.

The reports collected by Gallatin contain a number of explanations for the arrested development. Some were entirely personal. Nothing had been done on the Cape Fear Navigation after the death of the contractor, who had become the sole owner. On the Catawba and Wateree, the engineer had quit, "being hypochondriacally inclined." Few undertakings had professional engineers and several correspondents reported costly mistakes from lack of technical knowledge. The superintendents of one company, admitting many defects in the plan and execution of the work, explained in extenuation that they had "exerted themselves to the utmost and . . . never taken a single shilling" either for salary or expenses. But the most common explanation for the lack of progress ran in still simpler terms. "The principal difficulty . . . to be surmounted," reported the Appomattox Company, "arises from want of money." In case after case, work had ceased because of "the exhausted state of the funds." Several of the correspondents attempted to explain why the funds could not be replenished. In the case of the Dismal Swamp Canal, attention was called "to the peculiar circumstances of the inhabitants of the circumjacent country . . . to their possessing but little spare money, and their aversion to risk that little in speculation." Even in Philadelphia, it was pointed out that the funds of the moneyed interest were for the most part absorbed in investments promising "more advantage and immediate benefit." [11]

Although Gallatin's questionnaire did not ask for opinions on improvement policy, a number of his correspondents volunteered arguments for the use of federal funds to supply the deficiencies in the program. Some stressed the specifically national purposes of particular undertakings. It was argued, for example, that the Chesapeake and Delaware would give advantages to the United States Navy similar to those that the British expected from the Caledonian Canal through the Highlands of Scotland or—prophetically—that would someday be gained from a canal across the Isthmus of Panama. Others stressed the point, so often made by Washington, that development of means of communication with the settlers beyond the mountains was vital to the cementing and preservation of the Union. According to one correspondent, the best way to serve this patriotic purpose, and at the same time to "wither the consequence of Quebec," was to build a road uniting "the navigation of our great western lakes with that of the Atlantic." Combining the language of the classics with that of the Great North Woods, he described his project as "an Appian Way or national portage." Moses Brown, the Quaker manufacturer of Providence, advanced still another argument for federal action. Noting, as had many others, the intercity and interstate jealousies that had impeded the progress of improvements, he asked: "Would not such improving authority be proper to be vested in the General Government?" [12]

In one major case, that of the National Road, the general government had already assumed authority for internal improvement. This undertaking found its occasion in the settlement of the Northwest Territory and in the major political decision by which the public lands in the region were to remain at the disposition of the federal government. Even before the first state in the area was admitted to the Union, federal power over the land had been used to advance internal improvements. In an early example of the pragmatic arrangements linking public purpose with profitable private speculation, Ebenezer Zane, a pioneer entrepreneur of Wheeling, had marked out and cleared an important land route across southeastern Ohio (Zane's Trace) in return for the privilege of taking up town sites at the river crossings where he was certain that settlements would arise.[13] As Ohio's admission approached, improvement policy became a subject of more general discussion. Early in 1802, Gallatin proposed in a letter that part of the proceeds of the sale of public lands be used for the building of roads from "the navigable waters emptying into the Atlantic to the Ohio, and afterwards continued through the new State." Construction was to be carried on by the federal government with the consent of the states through which the road was to pass. He later endorsed his copy of the letter, "Origin of the National Road"; [14] it did in fact outline the policy which was followed. The Enabling Act of 1802, under which Ohio became a state, provided that five percent of the net proceeds of public land sold within its area should be applied to this purpose; and Congress later accepted the stipulation of the Ohio legislature that three fifths of the money should be spent on roads within the state.

The two percent fund began to accumulate, and late in 1805 a Senate committee reported in favor of building the road from Cumberland on the Potomac to Steubenville or Wheeling on the Ohio, crossing the mountains by the general route of the historic Braddock's Road. Considering it "an indelicacy" to assume that the Senate needed to be convinced of the importance of the undertaking, the Committee members devoted their main attention to the geographic advantages of the Cumberland route. Since this would provide direct connections for Baltimore and Washington, as against possible alternative routes from Philadelphia and Richmond, the choice was a controversial one and representatives from Pennsylvania and Virginia cast 29 of the 50 nays in a close vote in the House. In March, 1806, however, Congress adopted An Act to Regulate the Laying Out and Making a Road from Cumberland, in the State of Maryland. This authorized the President to appoint commissioners for laying out the road, to determine its location after receiving their report, and to request the necessary consent from the states concerned.[15] The Commissioners reported in December, noting as one of the difficulties of their task "the solicitude and importunities of the inhabitants of every part of the district, who severally considered their grounds entitled to a preference." [16] The state of Pennsylvania made its consent to the project contingent on the inclusion in the route of Uniontown and Washington, Pennsylvania. President Jefferson was somewhat disillusioned by this official expression of "solicitude" for local interests, but decided to accept the terms. Early in 1808, the President and the Secretary of the Treasury were able to report that the road had been located from Cumberland on the Potomac to Brownsville on the Monongahela, and that contracts had been let for clearing that portion of the route for half of its width.[17]

By this time, two other projects for internal improvements were under active discussion, the Chesapeake and Delaware Canal and the proposed canal around the Falls of the Ohio at Louisville. Each had been before the Congress since 1805, and each had received a favorable report from a Senate committee and an adverse report from a House committee on grounds of economy.[18] Both proposals were advocated on grounds of national interest. The advantages of a canal at Louisville would be felt throughout the entire area served by the Ohio River. The Chesapeake and Delaware was presented in the company's memorials as an essential part of a great system of inland navigation valuable to the nation in time of peace and still more vital in time of war. The Dismal Swamp Canal would carry the communication southward to Albemarle Sound, "and from thence through the inlets to South Carolina and Georgia." To the north, "a communication is now nearly opened from Albany, up the Mohawk River, to Lake Ontario and all the upper lakes," and the Hudson-Champlain canal would extend the navigation to Quebec. To the east, "the pass through Jersey" and a contemplated canal from Buzzard's Bay would "in like manner extend it to Boston and all the coast of Massachusetts." "Thus," the company concluded, "with opening only a few short passes, of which the Chesapeake and Delaware Canal is the great and preliminary one, a communication may be made, nearly free from all the dangers of the ocean, along the whole coast of the United States." [19]

Early in 1807, measures were before the Congress to aid these two enterprises by grants of western land in exchange for stock in the companies. The Falls of the Ohio bill was introduced in January by Henry Clay, and he had joined with the Delaware senators and other friends of internal improvements in active support of the two proposals. In the debate on the Chesapeake and Delaware project, John Quincy Adams, as his friend Senator Plumer reported, "was violent in his opposition." He spoke of the "league" of senators representing the six states immediately interested and warned of the fraud and collusion which would result from similar combinations in behalf of local measures.[20] After this, Adams introduced a resolution asking the Secretary of the Treasury to submit to the Senate a general plan of road, canal and river improvements which might require and deserve the aid of Congress—an action to which he pointed years later as marking the real beginning of systematic national policy.[21] The resolution was promptly defeated, but as the end of the session approached, it became evident to the improvement advocates that they would not obtain positive action on their measures. In this situation, the Senate passed a bill to appoint commissioners to investigate the Kentucky project, and Senator Thomas Worthington of Ohio introduced resolutions asking Treasury studies of the proposals for the Chesapeake and Delaware Canal and for a turnpike connecting the Atlantic states. Finally Worthington withdrew these two motions and presented instead the proposal for a more general study. This resolution, which followed closely the substance and language of the Adams proposal but which came from a man who had voted for specific improvement projects as consistently as Adams had voted against them,[22] won the support of a large majority. On March 2, 1807, the Senate directed the Secretary of the Treasury to prepare and report to its next session, together with information on the progress and prospects of existing improvements,

> a plan for the application of such means as are within the power of Congress, to the purposes of opening roads and making canals, together with a statement of the undertakings of that nature, which, as objects of public improvement, may require and deserve the aid of Government.[23]

Gallatin's Report on Roads and Canals, prepared in response to this resolution, reached the Senate on April 6, 1808. "The general utility of artificial roads and canals," it asserts, "is universally admitted." In some countries, "these improvements may often, in ordinary cases, be left to individual exertion, without any direct aid from Government." But in the American case, two major circumstances, "whilst they render the facility of communications throughout the United States an object of primary importance, naturally check the application of private capital and enterprise to improvements on a large scale." The first of these is the relative scarcity of capital. It is much more difficult than in Europe to attract investment by "prospects of remote and moderate profit." The second is "the extent of the territory compared to the population." With a sparse population, local traffic cannot be counted on to make an improvement profitable. In general, a canal will be unproductive unless it opens "a communication with a natural extensive navigation which will flow through that new channel." For this reason, "some works already executed are unprofitable; many more remain unattempted, because their ultimate productiveness depends on other improvements, too extensive or too distant to be embraced by the same individuals."

"The General Government," declares the Report, "can alone remove these obstacles." Its resources are "amply sufficient for the completion of every practicable improvement. . . . With these resources, and embracing the whole Union, it will complete on any given line all the improvements, however distant, which may be necessary to render the whole productive."

The argument continues:

The early and efficient aid of the *Federal* Government is recommended by still more important considerations. The inconveniences, complaints, and perhaps dangers, which may result from a vast extent of territory, can not otherwise be radically removed or prevented than by opening speedy and easy communications through all its parts. Good roads and canals will shorten distances, facilitate commercial and personal intercourse, and unite, by a still more intimate community of interests, the most remote quarters of the United States. No other single operation, within the power of Government, can more effectually tend to strengthen and perpetuate that Union which secures external independence, domestic peace, and internal liberty.

What, then, are the specific objects that on this argument require and justify action by the national government? Gallatin derives his answers from a broad view of the geography of the country. The main problems are to improve communications along the Atlantic seaboard and to create means of communication that will reach the settlers beyond the mountains and promote the development of the West.

With respect to the former, the United States possesses "a tide water inland navigation . . . which, from Massachusetts to the southern extremity of Georgia, is principally, if not solely, interrupted by four necks of land." The four are Cape Cod, New Jersey between the Raritan and Delaware rivers, the peninsula between the Delaware and Chesapeake bays, and the "marshy tract which divides the Chesapeake from Albemarle Sound." These should be cut by canals, which would total less than one hundred miles and would be useful "in peace or war" as protection against "storms and enemies." To this should be added "a great turnpike extending from Maine to Georgia . . . passing through all the principal seaports."

The problem of communication with the West presents a greater difficulty. From New York to southern Georgia, the way is blocked by the two great ranges of the Appalachians. "In the present state of science," it is useless to think of crossing them by canals. There are, however, places at which the upper waters of the western and Atlantic rivers are near enough together to make communication practicable. For each of four pairs of rivers, therefore—the Allegheny and the Susquehanna or the Juniata, the Monongahela and the Potomac, the Kanawha and the James, and the Tennessee and the Santee or the Savannah—the Report recommends improvements in navigation of the eastern river and a road connecting it with the western stream.

To the north, geography opens a different possibility. The Hudson is the only river which "breaks through or turns all the mountains." "The tide in no other . . . comes within thirty miles of the Blue Ridge, or Eastern chain of mountains. In the North river it breaks through the Blue Ridge at West Point, and ascends above the Eastern termination of the Catskill, or great Western chain." Advantage should be taken of this obvious gateway to open communication with the inland navigation of the Great Lakes.

Gallatin summarizes his proposals as follows:

I. From north to south, in a direction parallel to the seacoast,		
1. Canals opening an inland navigation for sea vessels from Massachusetts to North Carolina, being more than two-thirds of the Atlantic seacoast of the United States, and across all the principal capes, cape Fear excepted. . .	$3,000,000	
2. A great turnpike road from Maine to Georgia along the whole extent of the Atlantic seacoast.	4,800,000	
		$7,800,000
II. From east to west, forming communications across the mountains between the Atlantic and western rivers,		
1. Improvement of the navigation of four great Atlantic rivers, including canals parallel to them.	1,500,000	
2. Four firstrate turnpike roads from those rivers across the mountains, to the four corresponding western rivers.	2,800,000	
3. Canal around the falls of the Ohio.	300,000	
4. Improvement of roads to Detroit, St. Louis and New Orleans. . . .	200,000	
		4,800,000

III. In a northern and northwestwardly direction, forming inland navigations between the Atlantic seacoast, and the great lakes and the St. Lawrence,		
1. Inland navigation between the North river and Lake Champlain. . . .	800,000	
2. Great inland navigation opened the whole way by canals from the North river to Lake Ontario.	2,200,000	
3. Canal around the falls and rapids of Niagara, opening a sloop navigation from Lake Ontario to the upper lakes as far as the extremities of Lake Michigan.	1,000,000	
		4,000,000
Making together.		$16,600,000

These are national objects, and their benefits should "diffuse and increase the national wealth in a very general way." Yet the states through which these routes pass will receive more direct benefit than the others. Therefore, says Gallatin, "justice, and, perhaps, policy no less than justice," seem to require national aid to more local projects in other areas. For this purpose he adds $3,400,000 to bring his total to the round figure of $20,000,000.

How is this money to be raised and spent? "An annual appropriation of two millions of dollars would accomplish all these great objects in ten years." "In times of peace and under prosperous circumstances," this sum could be paid from the surplus revenues without taxation. Indeed, a still larger amount was being paid between 1801 and 1809 on the national debt, which would then be almost completely discharged. Or, as an alternative, the proceeds of the sale of public lands might be applied to this purpose. The improvements themselves would be of immediate and direct benefit to the purchasers of land, and "the United States, considered merely as owners of the soil . . . would be amply repaid" in the rise of value of the lands remaining unsold. If it were desired, also, the fund might be a revolving one. Projects once brought to profitable operation might be sold "to individuals or companies, and the proceeds applied to a new improvement."

As an essential first step, the Report suggests that Congress should authorize taking "the surveys and levels" of the routes proposed. In the present constitutional position, the federal government can make improvements only with the consent of the states on whose territory they are to be built. For this reason, President Jefferson has proposed the adoption of an amendment "to remove every impediment to a national plan of internal improvements." Yet the specific projects put forward in the Report are, Gallatin argues, of such obvious advantage that the consent of the states should be easy to obtain. As for the method of organization, the United States may have the work done "at their sole expense," or may subsidize private companies, preferably by subscription to their stock. The former method, "by effectually controlling local interests," would perhaps assure better general direction. The latter might be more economical in detailed execution. Perhaps the two might somehow be "blended" to obtain the advantages of each. In the closing paragraph the Secretary returns to the problem of local interests in the choice of projects in a sentence notably optimistic in its choice of conjunction: "The National Legislature alone, embracing every local interest, and superior to every local consideration, is competent to the selection of such national objects."

Here, then, was a notable ten-year plan of national action. The economic argument is modern in substance if not in phraseology. The specific recommendations were based, as the hundred and eighty folio pages of appendix indicate, on a painstaking investigation of improvements already undertaken or projected. Most of the elements, indeed, were already familiar. Gallatin's four "necks" corresponded to the four "passes" of the Chesapeake and Delaware enthusiasts. The great turnpike was the main post road. Improvements had been started on the Hudson, the Mohawk, the Potomac, and the James. One of Gallatin's four highways connecting eastern and western rivers, the Cumberland Road, was already being built under his own direction. What distinguishes the Report from earlier discussion is not the novelty of its particular proposals, but the presentation of a coherent program of projects of national importance. Based, as Gallatin said, on the "great geographical features of the country," their selection demonstrated his keen appreciation of the natural advantages of the United States and the ways in which they might be shaped to human use.[24]

The principal improvements proposed in the Report are presented in the map on p. 34 and compared with later construction on the same or similar routes. The four canals along the seaboard were all finally constructed, and today's Inland Waterway as well as highway and railway connections follow the general direction of Gallatin's proposals. On the more significant problem of connection with the West, the map indicates that—subject to certain deviations—the Erie Canal and four of the five trunk-line railroads all took advantage of the openings suggested in the Report. Though changes in technology and economic circumstance reduce the relevance of the criterion of later development, the comparison tends to corroborate the firmness of Gallatin's grasp of geographic reality.

The political and administrative aspects of the plan, on the other hand, are much less fully developed. The basis for selecting objects of truly national importance is indicated by example rather than precept. The alternate forms under which federal action might be organized are discussed only briefly; and the choice between them is put forward, perhaps for tactical reasons, more as a matter of detail than of substance. The constitutional issue is presented as offering no great difficulty. The question of conflicting state and local interests, though clearly seen, is answered with a hopeful phrase and the "justice and policy" fund for appeasing the states less favored by the major projects. Yet it was on these points that the principal difficulties were to appear.

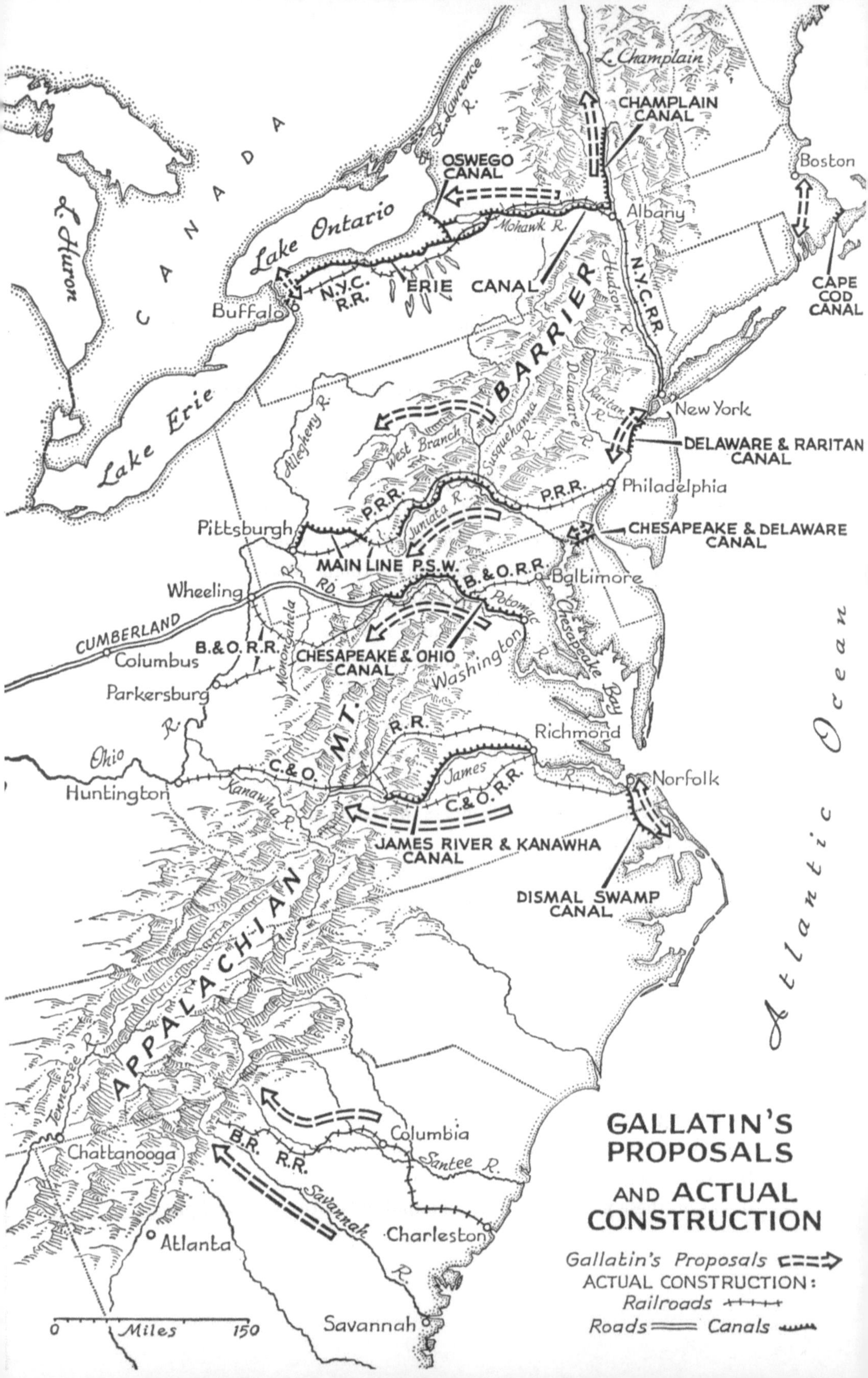

GALLATIN'S PROPOSALS AND ACTUAL CONSTRUCTION
Gallatin's Proposals
ACTUAL CONSTRUCTION:
Railroads
Roads
Canals
0 Miles 150
L. Champlain
CHAMPLAIN CANAL
St. Lawrence R.
OSWEGO CANAL
CANADA
L. Huron
Lake Ontario
Mohawk R.
Albany
Boston
CAPE COD CANAL
N.Y.C. R.R.
ERIE CANAL
Buffalo
BARRIER
Hudson R.
N.Y.C.R.R.
Lake Erie
Allegheny R.
West Branch
Susquehanna R.
Delaware R.
Raritan R.
New York
DELAWARE & RARITAN CANAL
P.R.R.
P.R.R.
Philadelphia
Juniata R.
Pittsburgh
MAIN LINE P.S.W.
CHESAPEAKE & DELAWARE CANAL
B.&O.R.R.
Baltimore
Wheeling
RD.
CUMBERLAND
Potomac R.
Chesapeake Bay
B.&O.R.R.
Columbus
Monongahela R.
CHESAPEAKE & OHIO CANAL
Washington
Parkersburg
MT.
R.R.
Richmond
Ohio R.
C.&O.
James R.
Norfolk
Huntington
Kanawha R.
C.&O.R.R.
JAMES RIVER & KANAWHA CANAL
DISMAL SWAMP CANAL
Atlantic Ocean
APPALACHIAN
Tennessee R.
Columbia
Chattanooga
B.R. R.R.
Santee R.
Savannah R.
Charleston
Atlanta
Savannah

GALLATIN'S PROPOSALS AND ACTUAL CONSTRUCTION

Gallatin's proposals	*Actual construction*	*Agency*	*Year completed*	*Year abandoned*
The Four Necks of Land				
(1) Cape Cod	Cape Cod Canal	Private	1914	—
(2) New Jersey	Delaware and Raritan Canal	Private	1834	1933
(3) Delaware and Chesapeake Bay	Chesapeake and Delaware Canal	Mixed (federal, state, private)	1830	—
(4) Chesapeake Bay to Albemarle Sound	Dismal Swamp Canal	Mixed (federal, state, private)	1828	—
The Northern Opening				
Hudson to Lake Champlain	Champlain Canal	State	1823	—
Mohawk to Lakes Ontario and Erie	Erie Canal	State	1825	—
	Oswego Canal	State	1828	—
	New York Central Railroad	Mainly private [a]	1841–1851 [d]	—
The Four Pairs of Rivers				
(1) Allegheny-Juniata, Susquehanna	Main Line—Pennsylvania State Works	State	1834	
	Philadelphia & Columbia R.R.			—
	Susquehanna and Juniata Divisions (Canal)			1901
	Portage Railroad			1857
	Western Division Canal			1865
	Pennsylvania Railroad	Mixed (local government, private) [b]	1855 [e]	—
(2) Monongahela-Potomac	National (Cumberland) Road	Federal Government	1818 [e]	? [g]
	Chesapeake & Ohio Canal	Mixed (federal, state, local government, private)	1850	1924
	Baltimore & Ohio Railroad	Mixed (state, local government, private)	1853 [e]	—
(3) Kanawha-James	James River & Kanawha Canal	Mixed (state, local government, private)	1851	1880
	Chesapeake & Ohio Railroad	Mixed (state, local government, private) [c]	1873 [e]	—
(4) Tennessee-Savannah, Santee	Blue Ridge Railroad	Mixed (state, local government, private)	1861 [f]	—

[a] Local aid between Albany and Schenectady.

[b] Philadelphia & Columbia section purchased from state.

[c] Two of its components, the Blue Ridge Railroad (of Virginia) and the Covington & Ohio, were built as state works. The state's interest was sold to private purchasers before the railroad was completed to the Ohio River.

[d] 1841, Albany to Buffalo. 1851, Hudson River Railroad from New York City. New York Central Railroad formed by consolidation in 1853. River at Albany not bridged until 1866.

[e] To Pittsburgh; Wheeling; Wheeling; and Huntington, W. Va. respectively.

[f] Further work abandoned.

[g] No significant date can be assigned. Different sections of the road fell into disuse at different times. The present U.S. Route 40 follows much of its course.

One major political obstacle was evident even before the Report was submitted. Its fiscal proposals, which contemplated neither new taxation nor borrowing, were based explicitly on the maintenance of peace, the continuation of substantial revenue from customs, and the absence of great military expenditure. But the Embargo Act had been passed in December, 1807; and the years that followed were characterized by interruptions to trade, by the creation of a military establishment, and finally by the War of 1812. Even under these circumstances, the advocates of internal improvement, armed with numerous petitions on behalf of particular projects, made various attempts to persuade the Congress that "foreign relations ought not to exclude every other object of legislation." A small appropriation for the Carondelet Canal at New Orleans was made in 1809 as a defense measure. Senator John Pope of Kentucky and Representative Paul B. Porter of New York introduced in 1810 a comprehensive bill to carry out a list of projects similar to Gallatin's, choosing the method of subscription to the stock of mixed enterprises and proposing to finance the contribution by borrowing on the security of tracts of public land.[25] Bills for the Chesapeake and Delaware and Falls of the Ohio canals several times won Senate majorities.[26] In 1811, Gouverneur Morris and De Witt Clinton were sent by New York State's Canal Commission to appeal for federal support for the proposed canal to the Great Lakes. In an attempt to win the widest possible support, they inspired the drafting of an extraordinary omnibus bill which was to set aside all the land in Michigan Territory north of the fortieth parallel and give it, for the purposes of various improvements, to twelve states and the Potomac Company, with New York receiving 4,500,000 acres out of a total of just under 10,000,000 acres.[27] But early in 1812, a House committee appointed to consider these and other projects drew from Gallatin himself the emphatic statement that direct appropriations could not be considered in the present state of the nation's finance, and the committee reported with regret its opinion "that the inauspicious situation of the United States, in regard to our foreign relations" rendered it "inexpedient for Congress of the United States to make a donation in land or money." [28] The committee's policy prevailed, and its arguments were reinforced by the outbreak of war.

The return of peace removed these obstacles and revived the agitation for internal improvement. The war experience itself furnished an additional argument. The effectiveness of the navy had suffered for want of internal navigation along the seacoast, and that of the army by lack of proper transport to the Great Lakes. The chartering of the Second Bank of the United States seemed to provide a special opportunity for action, and John C. Calhoun appeared as its leading advocate. During the session of 1816–17, he proposed a measure to set aside the federal government's bonus from the charter and its future income from the Bank as a fund for internal improvements. "Let us then, said Mr. C."—as the *Annals of Congress* reported—"bind the Republic together with a perfect system of roads and canals." For this the state of the nation was now most favorable, "at peace with all the world, abounding in pecuniary means." To these two circumstances, Calhoun added a third "of the most importance . . . party and sectional feelings immerged in a liberal and enlightened regard to the general concerns of the nation." [29]

In spite of this optimism, the conflict of state versus more general interests appeared at once in the Bonus Bill discussions. As introduced, the measure provided for a permanent national fund to be spent under the direction of Congress for improvements of national importance. As passed, it represented hardly more than a distribution of gifts to the states in proportion to their population. The fund was first to be carved into portions for the several states and then within each state applied to improvements under the concurrent authority of Congress and the legislature. Clay and Calhoun and others pointed out that this would make it very difficult to secure the construction of projects of national importance, citing as a favorite example the canal around the Falls of the Ohio which would be of more benefit to states up and down the river than to Kentucky. Yet they could secure only the proviso that a state might, if it wished, ask Congress to spend its share of the fund within the borders of another state.[30] In the end the bill failed because of President Madison's veto; but the measure, which he considered beyond the powers of the federal government, would in any case have given it too little authority for the systematic national planning called for in Gallatin's Report. The same limitation, moreover, was acceptable to most of those who tried, unsuccessfully, to clarify the government's powers over internal improvements by the process of constitutional amendment.[31]

Meanwhile the National Road, on which construction had continued even during the war, reached the Ohio at Wheeling in 1818 and began at once to carry substantial traffic.[32] The acts admitting Indiana, Illinois, and Missouri to statehood all contained provisions similar to those of the Ohio Act and seemed to assure its extension. But this enterprise also, as its historian remarked, "swung . . . on a swaying constitutional rope."[33] In 1822, Monroe vetoed a measure authorizing the government to establish toll gates and collect revenue for its maintenance. Without such power there could be little prospect of its continuance and operation as a fully national work.

New projects were slow to win support in spite of the favorable circumstances cited by Calhoun. As Secretary of War, he presented in 1819 a comprehensive plan of improvements, similar to Gallatin's, but added projects further to the west, and buttressed it with forceful arguments on both military and developmental grounds. A House committee in 1822 approved a bill embodying these recommendations. An important step toward the more orderly consideration of improvement projects was taken in the General Survey Act of 1824, which instructed the President to arrange for surveys of the routes of canals and roads of national importance.[34] As a result, the reports of the United States Corps of Engineers and its Board of Internal Improvement began to play an important part in the discussions in the House and Senate and in their respective committees on Roads and Canals. The persistent efforts of the sponsors of the Chesapeake and Delaware Canal won their first success early in 1825, and a subscription to its stock was approved by Monroe on his last day in office.

The inauguration of John Quincy Adams appeared to be the sign for more vigorous action. His first annual message asked the Congress to consider "the general principle" of internal improvement "in a more enlarged extent." The federal government should carry out "works important to the whole . . . and to which neither the authority nor the resources of any one State can be adequate." Impartial and scientific judgment in the selection of objects would be provided by the Board of Engineers for Internal Improvement. "The swelling tide of wealth" from the sale of public lands would be "made to reflow in unfailing streams of improvement from the Atlantic to the Pacific Ocean." [35] With the inclusion of the ideas of the great revolving fund and of the planning agency, this may indeed be taken as the most "enlarged" expression of the philosophy of internal improvements as a national system.

Charles F. Mercer, who was chairman of the House Committee on Roads and Canals and already well known for his part in creating the Virginia system of internal improvements, introduced a set of resolutions which proposed to set aside a fund for internal improvements and to devise a system of rules for its impartial and judicious application. To aid Congress in the selection of objects an enlarged Civil Engineer Corps was to be set up within the Army Engineers to "gather information on the susceptibility of the United States to improvement by roads, canals and railways" and also to supervise the work itself. The progress of all internal improvement was to be recorded in the Census.[36]

Positive action took the form of aid to specific projects. One method adopted was that of land grants to individual states: to Ohio for the construction of a turnpike from Columbus to Sandusky on Lake Erie; to Indiana for a road the length of the state from Lake Michigan to the Ohio River, and also for a canal from the Wabash River to Lake Erie; and to Illinois for a canal to connect the Illinois River with Lake Michigan. All these could be thought of as westward extensions of the Gallatin plan, linking the commerce of the Lakes and the Erie Canal with that of the Mississippi system; and the proposal to make the short connection at Chicago had long been the subject of congressional eloquence on the creation of a "circumnavigation" of the United States.[37] Three of these measures set the important precedent of granting alternate sections of land in strips along the improvement, with the remaining half reserved for the United States.[38]

Federal public works were authorized by the passage in 1824 of the first river-improvement bill and the first harbor-improvement bill, the latter following earlier precedents in lighthouse construction. Two years later the Congress adopted the first of the long series of combined "rivers and harbors" measures.

Principal reliance, however, was placed on the method of subscription to the stock of improvement companies. This was generally accepted as meeting the constitutional difficulty, since the company's powers were derived from a charter granted by the state. This method, moreover, appeared to require smaller appropriations; and the prevailing opinion was that the government's money would be spent more wisely if the judgment and zeal of the private investor were also involved. In this way, as a Senate committee had argued, the government was less likely to "become engaged in impracticable projects for improvement" and more certain of "the economical expenditure of the funds."[39] A second subscription was made to the stock of the Chesapeake and Delaware Canal, and two subscriptions each were made to the Dismal Swamp Canal Company and to the Louisville and Portland Company which was to make the improvement at the Falls of the Ohio. In May, 1828, Congress appropriated a million dollars for the greatest of the proposed improvements—the

Chesapeake and Ohio Canal—the heir of Washington's Potomac Company. In this case, the federal government was the largest stockholder, and state and municipal subscriptions outweighed those of private investors.[40] The President took great satisfaction in turning the first earth for the canal on the Fourth of July.[41]

By the close of Adams' administration these four undertakings, each a part of Gallatin's original plan, had received the following appropriations:

Chesapeake and Delaware Canal Company	$300,000	Approved	March 3, 1824
	150,000	"	March 2, 1829
Louisville and Portland Canal Company	100,000	"	May 13, 1826
	135,000	"	March 2, 1829
Dismal Swamp Canal Company	150,000	"	May 18, 1826
	50,000	"	March 2, 1829
Chesapeake and Ohio Canal Company	1,000,000	"	May 24, 1828

The three smaller canals were almost completed, the large one just begun. The National Road had been built across the mountains and into Ohio, but the federal government could not collect tolls and operate it as a turnpike. Something more than two million dollars for the Road, less than two million in stock subscriptions, and land grants which were ultimately to amount to some 2,500,000 acres—these had been the tangible contributions of the federal government to internal improvements.

The prospects were decisively changed by the election of Andrew Jackson to the presidency. The principal announcement of new policy was the famous Maysville Road veto of May 27, 1830. The measure proposed a subscription of $150,000 to the stock of a company which was to build a sixty-mile turnpike. Though it was to continue the line of Zane's Trace, the road itself was to be wholly within the single state of Kentucky. It was therefore easy for the President to describe it as "a measure of purely local character . . . conferring partial instead of general advantages" and bearing "no relation to any general system of improvement." The organization of the project, moreover, made it vulnerable to the charge that it represented an expedient for saddling upon the government "the losses of unsuccessful private speculation." It may be said that Jackson was in effect condemning the Maysville project precisely on the ground that it did not meet the specifications laid down by Gallatin. But projects which did meet these specifications fared no better. A further appropriation for the Louisville and Portland Canal received a pocket veto.[42] The federal government withdrew from leadership in the affairs of the Chesapeake and Ohio Canal Company, and Congress refused all its petitions for further aid.[43] Appropriations continued for the extension of the National Road; but its completed portions were handed over to the states of Ohio, Pennsylvania, Maryland, and Virginia, which could exercise the powers of operation and maintenance denied to the federal government under the prevailing constitutional doctrine. Referring to internal improvements in his Farewell Address, Jackson boasted that he had "finally overthrown . . . this plan of unconstitutional expenditure for the purpose of corrupt influence";[44] and the Era of National Projects may be thought of as ending in his administration.

This does not mean that all forms of federal aid to transportation were completely abandoned. The National Road continued to be built westward at national expense until the mid-century, reaching Vandalia, then the capital of Illinois, but it never extended to the Missouri border. Federal appropriations for construction and maintenance of the seven-hundred-mile highway reached a total of nearly $7,000,000;[45] but its character as a fully national improvement had been lost, and the final sections were ultimately ceded to Indiana and Illinois. Military roads built through the territories contributed to economic development as well as to defense. Appropriations for rivers and harbors improvement increased until 1838 and continued intermittently after that date.[46] The Army Engineers expanded their activities not as central planners but as a service agency for these federal projects and also for many state and private undertakings. As Forest Hill has shown, they contributed to the transportation development of the time a large share of the scarce and indispensable factor of engineering skill.[47]

Yet the negative decisions, though never officially stated in precise terms, were in their broad lines unmistakable. The national government might aid transportation in various ways which returned no direct income, but it must refrain from building revenue-producing public works. It was not to construct roads and canals on which tolls were to be collected. It was not to subscribe to the stock of improvement companies. Most certainly, it was not to undertake the burden and the responsibility of a comprehensive and scientifically planned system of national improvements.

"Posterity," said one congressional debater, would find it difficult to "credit" or understand the success of the opponents of federal action, especially considering "the enlightened period in which we live, and the spirit of universal improvement." [48] Why, then, had the government carried out so small a part of Gallatin's ten-year plan in the twenty years between its presentation and the election of Jackson? And why had the opponents of the program then prevailed?

The explanation clearly does not lie in the competition between public and private enterprise. The issue as debated in these early decades of the Republic bears little analogy to the present-day controversy as to whether the federal government or a great utility company should develop the hydroelectric power of Hell's Canyon. Most participants in the debate took it for granted that it was desirable to leave to private enterprise the things which it could accomplish, but few or none believed that this included major improvements like those comprised in the Gallatin plan. Though many considered that the willingness of private investors to share in the cost was a good criterion for the soundness of government-aided projects, it was only rarely argued that the works themselves should be put off until "industrious and careful people" were ready to build them entirely with their own money "for a reasonable profit." [49] The basic individualism of the country had its most direct effect on the federal improvements program through the reluctance of both the leaders and the people to sanction substantial taxation. Appropriations therefore were hard to secure, and the same factor had some influence on the choice of the method of action which was subject to attack as "stock-jobbing" and as a degrading association between the government and groups of its own citizens. On this point Jacksonian opposition was based on a desire to keep business out of government—if the modern phrase can be applied—rather than a desire to keep government out of business. Certainly the small and struggling road and canal companies of the time were more likely to be found petitioning for federal aid, as in the Maysville case, than complaining against federal competition.[50]

The real issue was between national and state action. This was clear in the constitutional debates which filled so many pages in the *Annals of Congress* and in the works of later historians. In these the issue turned solely on the powers given by the Constitution to the federal and state governments respectively and not at all on the rights of private individuals and corporations. Advocates of federal action made elaborate attempts to find justification in various powers enumerated in the document, and on many occasions presidents or legislators professed themselves to be in sympathy with the purposes of the measures proposed but unable to accept the arguments for their constitutionality. Behind the subtleties of interpretation there was no doubt a deep-seated reluctance to strengthen the powers of the general government and to increase the amount of "executive patronage." Modification or looser construction of the constitution in one field was sometimes opposed for fear that it might make it easier to extend federal powers in other fields and even—as was occasionally suggested—to such an extreme as the emancipation of the slaves.[51] Yet it is difficult for a modern student to believe that the constitutional issue, for all the passion with which it was contested, was in fact the decisive one. For many years the method of stock subscription satisfied the constitutional scruples of most of the legislators. The right to go farther and to create a fully national system of improvements could have been secured by constitutional amendment. If there had been a great positive demand for a federal program, it would be hard to understand why more was not accomplished under the former method and why the advocates of improvement did not make a more serious attempt to amend the constitution.

The lack of compelling popular pressure on behalf of a national program was in part due to a belief that the states were themselves well equipped to carry out the essential improvements. Jackson argued that public funds for the purpose would be "more judiciously applied and economically expended under the direction of the State legislatures." [52] The advantages of state action were described by the enthusiasts who contributed articles on internal improvements to the *North American Review:*

> The compass of each state is sufficiently narrow, and its legislative power sufficiently diffused, to render knowledge of its internal condition, wants, and resources easily attained. . . . Scarce any object of public utility is beyond the grasp of the resources of the single states; so that, after all, the care of individual objects of public improvement is put into the hands of those most sure to be benefited by them.[53]

These arguments were used again and again by congressional orators, and acquired particular force when it became possible to compare New York's success with the Erie Canal with the federal government's slow progress with the Cumberland Road.[54]

In the defeat of the federal program, however, the conflicts of state and sectional interests seem to have played a larger part than either the belief in states' rights or any positive faith in the states' capacities. Throughout the debates these rivalries stood in the way of the adoption of specific measures or agreement on a general program. One Maryland congressman declared that Pennsylvania seemed more inclined "to put a mountain in the middle of the Cumberland road than to repair it," and another saw his enemies in another direction when he complained that scarcely a single vote from the "southward and westward of the Potomac" had been cast in favor of the Chesapeake and

Delaware Canal.[55] More frequently, internal improvements appeared as a demand of the West. They were often urged as an aid to western agriculture, "cheapening the carriage of its heavy products," and as an offset to the favors already given to eastern commerce and manufactures. There was even, warned Congressman Porter, danger of revolt if the people of the West came to believe that Congress "can constitutionally create banks for the accommodation of the merchant but cannot construct canals for the benefit of the farmer." [56]

Still another alignment divided states which had already spent substantial sums on improvement from those which had not done so. The New England states belonged to the former group. New York and Pennsylvania, on main routes to the West, were at first eager for federal assistance and willing to support broad programs of action. After their demands had been rejected, and they had built their own improvements at heavy cost, their representatives were understandably more reluctant to vote appropriations for the projects of other states.[57]

Adams's charge that the national program was overthrown by what he called "the Sable Genius of the South" is therefore a great over-simplification. There were other cleavages besides that between North and South to make agreement difficult. Calhoun in his War Department days and Mercer of Virginia were as strong adherents of a fully national system of improvements as were Gallatin of western Pennsylvania or Adams of Massachusetts or Clay of the border state of Kentucky. Yet as the debate continued, the most implacable opposition to a national program came from the South. For this there were obvious explanations both in the geographic setting and in alignment of political forces. The greatest of the improvements, in Gallatin's Report and in subsequent action, were those designed to overcome the Appalachian Barrier. Their function was to provide trade connections between East and West. They brought little benefit to the southern states and to the Lower South, none at all. Without such intervention, on the other hand, the rivers would serve to carry western produce to the South and southern cotton to the sea.

To the political explanation, Adams's own phrases provide the clue. He speaks of the South as "cursing the tariff and internal improvements." [58] Protection to manufactures was clearly inimical to the interests of the South as a region exporting agricultural products. The tariff was linked to internal improvements as a principal source of the funds from which they might be constructed and—more important—as an element in the American System of Henry Clay which offered tariff protection to the East and internal improvements to the West. Since success in the creation of an East-West alliance on such a basis would obviously threaten the political position and influence of the slaveholding states, it is not surprising to find a stiffening of southern opposition to both elements in the program.

Resort to national action had been advocated by those who feared that state and local conflicts of interest would prevent the construction of needed improvements. The effect, they thought, would be to place decisions in the body "in whose councils a local or partial spirit is least likely to predominate." [59] But in the Congress, state and regional differences were from the beginning too great to permit agreement either on a broad national program or on a large number of individual projects. As the lines of sectional controversy hardened, the possibility of such agreement became still more remote.[60]

In 1837, lamenting the fall of the program of "internal improvements by means of the national energies," Adams wrote:

> With this system in ten years from this day the surface of the whole Union would have been checkered over with railroads and canals. It may still be done half a century later and with the limping gait of State legislature and private adventure.[61]

Even by the close of the Era of National Projects, "private adventure" was hardly strong enough to claim more than a junior partnership in the larger projects. In default of national action, leadership in internal improvements passed to the states and in some cases to city governments. It is therefore necessary to turn to the field of state and local action to see what progress, "limping" or otherwise, was made in breaching the Appalachian Barrier and in building the essential network of transportation.

THE EMULATION OF STATES AND CITIES

1815-1861

. . . an example to other states worthy of emulating enterprise.

James Madison on the completion of the Erie Canal, 1825

The true spirit of Internal Improvement is abroad in the *West.*

American Railroad Journal, 1836

3

THE APPALACHIAN COMPETITION

With the federal government largely out of the field, the emulation of states and cities became the most conspicuous feature of the movement for internal improvement. The advocates of roads and canals, and soon of railroads, pushed their projects in every state and in almost every locality. Some of these followed coastal routes, many were of narrowly local character, and others were directed toward the opening of the West. These last were of the greatest general interest, and until the middle of the nineteenth century the leading object of internal improvement continued to be—as in the Gallatin Report—the passage or the turning of the Appalachian Barrier. The direct contributions of the national government toward this goal had been the construction of the Cumberland Road and the million-dollar subscription to the Chesapeake and Ohio Canal Company. State efforts to overcome the barrier had begun before the federal government withdrew, and the work continued under a great variety of state, local, and private auspices.

In the competition for "that interesting prize," the commerce of the West,[1] attempts were made to develop each of the five routes described by Gallatin: the opening in the North provided by the valleys of the Hudson and the Mohawk and the four pairs of eastern and western rivers. The first route belonged to New York State and the Susquehanna-Allegheny connection to Pennsylvania. The Potomac gateway, through which both the federal projects had been directed, became principally the concern of the state of Maryland and the city of Baltimore. The James and Kanawha Rivers lay within the boundaries of Virginia. The breaching of the barrier of the Southern Appalachians to reach the upper waters of the Tennessee was the objective of ambitious projects of South Carolina and the city of Charleston, and to a lesser degree of North Carolina. The alternative turning of the barrier at its southern end was to be the accomplishment of the state of Georgia.

The first and greatest success in these competing efforts was made by the northernmost of these routes. Even before 1800, the Western Inland Lock Navigation Company, following the old Oswego route of the fur traders, had accomplished an imperfect connection between the Mohawk River and streams flowing into Lake Ontario. The company, however, had long since run out of funds, and the undertaking which replaced it differed in three major respects. In the first place, it was to be a fully independent canal rather than a set of river improvements with short canals at the major obstacles. In the second place, it was to be a single "grand canal" extending all the way to Lake Erie instead of two much shorter canals, one to Lake Ontario and the other around Niagara, as projected in the Gallatin Report. Though this innovation seemed so shocking to the 1808 Legislature that it was unwilling to give explicit approval even for a survey of the proposed route, and though many observers agreed with President Jefferson that the idea of a canal extending three hundred and fifty miles through the wilderness was a century ahead of its time, the final decision was the bold one, to strike directly to Lake Erie to win the trade of the upper lakes and the American West.[2]

The third decision was that an undertaking of such magnitude must be carried on as a public work rather than by private initiative with state aid and participation. It was at first not clear that this would be by New York State alone. Canal commissioners appointed under an act of 1810 attempted to enlist the cooperation of the federal government and of other states that would benefit from the canal. From the states they received nothing more tangible than resolutions of encouragement. Of federal aid, they had high hopes on two occasions—on the eve of the War of 1812 and again at the time of the Bonus Bill, which would have brought to the state $1,500,000. The Bill was vetoed by President Madison on March 3, 1817. On April 15, the New York Legislature, acting, it was said, "under the scoff and hiss of [the] general government," [3] adopted a measure providing for the building of the canal on the sole responsibility of the state.

This act authorized the Canal Commissioners to begin the construction of the middle portion of the Erie Canal, from the Mohawk to the Seneca River, and also of the Champlain Canal, following the route of the old Northern Inland Lock Company, from the Hudson to Lake Champlain. As the basis for these and subsequent efforts, it established a permanent Canal Fund which was to obtain its income from three main sources: first, by the earmarking of certain state revenues (taxes on salt manufacture and steamship passages, and part of the proceeds of lotteries and auction duties); second, by borrowing on the credit of the state; and, eventually by the collection of tolls on the canals themselves.[4]

The first earth was turned on July 4, 1817. On the 4th of November, 1825, Governor De Witt Clinton, who had from 1810 to 1824 been the leading spirit among the Canal Commissioners, celebrated the opening of the canal by pouring into the Atlantic a keg of water from Lake Erie. The historian of the celebration summarized the achievement in the following words, setting them in full capitals: "They have built the longest canal in the world in the least time, with the least experience, for the least money, and to the greatest public benefit." [5] If one may disregard the canals of ancient China, these superlatives are a fairly sober measure of what had been accomplished by the agents of a state whose population was little more than a million when the work began. Without professionally trained engineers—Benjamin Wright and James Geddes, the two principal engineers had each been lawyer, judge, and surveyor—they had devised methods of canal-building and contracting that were to be followed in many other undertakings. They had constructed 363 miles of canal, much of it through unsettled country, for a little more than $7,000,000. The raising of this money itself involved innovation. Borrowings during the early years were relatively small. As Nathan Miller has shown, subscriptions to the first loans were made almost entirely by citizens of the state and for the most part in modest amounts. But when the financial success of the canal was foreshadowed by the collection of substantial tolls as soon as the middle section was opened for traffic, the securities became attractive to larger investors in New York City and then in London. The bigger issues of the later years were largely marketed abroad, and by 1829 more than half of the outstanding canal debt was held by foreigners.[6]

As for "public benefit," part of the evidence lies in the financial returns. The $7,000,000 canal collected $1,000,000 in tolls before it was open for passage from end to end. From the time of completion tonnage on the canal began to rival that of the entire Mississippi system. Operation showed a substantial profit from the beginning. By 1836, after two reductions in the rate of tolls, the surplus from the Canal Fund was great enough to provide for payment of the entire debt incurred for the construction of the Erie and Champlain Canals. Half of this surplus, to be sure, had come from the contributions of the salt and auction duties,[7] but the annual excess of tolls over expenditures on the two canals was now over $1,000,000. From this time on these tax revenues reverted to the general fund, and beginning in 1835 the legislature made a substantial levy on the canal tolls for the general purposes of the state.

Still more important was the effect on economic development. The force of the impact may be suggested by the often-cited fact that the cost of shipping a ton of wheat or flour from Buffalo to New York City fell at once from one hundred to ten or twelve dollars. Now for the first time settlers in the American West could ship heavy and bulky produce overland to the East rather than down the Mississippi or perhaps the St. Lawrence, and the basis was laid for the increasingly significant exchange of eastern manufactures for the products of western agriculture. The stimulus to the commerce of New York City, to the growth of cities along the canal route, and to the spectacular development of the old Northwest, as well as the canal's service as a great channel of westward migration, have been often noted.[8] Its opening, indeed, may be regarded as the most decisive single event in the history of American transportation.

But the story of the New York State's public effort in improvement by no means ends with the construction of the original Erie Canal, and some of the later chapters are much less triumphant. An immediate effect of the Erie's success was to stimulate public demand for more and more canals. A law of 1825 authorized surveys for no less than seventeen other routes. The Champlain Canal had been completed two years before the Erie, at a cost of about $1,000,000, and was doing a profitable business. The branch to Lake Ontario, by the Oswego Canal, was finished in 1828 and attracted substantial tonnage, which some thought was at the expense of the western division of the Erie itself. In the same years the Cayuga and Seneca Canal connected these two lakes with the main system, and five years later the Crooked Lake Canal extended the connection to Lake Keuka. During the thirties two southern laterals were constructed, each extending to the waters of the Susquehanna and intended to connect with the Pennsylvania canal system—the Chemung from Lake Seneca to Elmira, and the Chenango from Utica to Binghamton. Toward the end of the decade, work was started on a third southern lateral, the Genesee Valley, to be built from Rochester to Olean on the Allegheny, and on a northern lateral, the Black River, to serve the farming country west of the Adirondacks and the timber trade of the mountains and to bring a supply of water to the main canal. In addition, it had become evident that the Erie Canal itself must be made larger and more efficient in order to handle the continual increases in the volume of traffic, and to meet the competition from other states "and especially the provinces of Canada." [9] A program of enlargement, which eventually was to cost more than four times the original investment, was accordingly authorized by the legislature in 1835.

Nor were these the only efforts toward internal improvement. The Delaware and Hudson Canal, projected by a private company to carry anthracite from the Pennsylvania coal fields to the Hudson at Kingston, was completed in 1829 with the aid of state loans of $800,000 and became highly profitable. Assistance on a similar basis was given to a number of railroad companies. Little of this went to the series of short lines paralleling the "water-level route" of the Erie Canal; indeed, the state for some time acted to protect its own enterprise from their competition by forbidding the carriage of freight or permitting it only during the closed season of the Canal and on payment of canal tolls.[10] Instead, the largest amount of aid went to the New York and Erie Railroad, which was to cross the state by a shorter route but over rougher terrain through a region almost devoid of major settlements. In this case the arguments for assistance ran largely in terms of justice to the people of a region which received little benefit from the canal system of the state. A loan of $3,000,000 was made under legislation of 1836 and 1839. In the late thirties and early forties, the state also aided ten other railroad companies to the extent of a little more than $2,000,000. In these cases its experience ranged from that of the Catskill and Canajoharie, which never built its road and cost the state more than $350,000, to those of six railroads whose loans were promptly repaid.[11]

The expansion of this program received an abrupt check at the beginning of the forties. The state was in financial difficulty and its improvement securities, which before had commanded premiums, could not be sold except at a discount. The $3,000,000 loan for the Erie Railroad yielded the company less than $2,600,000. This was partly because of the depression following the crisis of 1839 and partly because of the amount of expenditures and commitments that had already been made. The Oswego Canal and the other early branch and lateral canals had cost more than $3,500,000. The railroad commitments were another $5,000,000. It was estimated in 1840 that three works already authorized and begun, the Erie Canal enlargement and the Genesee and Black River laterals, would cost no less than $30,000,000. The state debt, which was under $7,000,000 in 1837, rose to nearly $27,000,000 in 1842. In the meanwhile, faith in the prospects of the canal system, formerly so unbounded, had been weakened by the formidable competition of the railroads. On the other hand, the activities of the lobbyists for the railroad companies, whom one politician described as "asking the legislature to give them leave to thrust their hands up to the elbows into the public treasury,"[12] provided an argument against further aid to railroads. In 1840 Governor Seward pointed out that continuation of canal expenditures during the depression had aided the economy of the state; but he argued that any new loans should be based solely, as before, on "the existing and anticipated revenues of the canals . . . since taxation for internal improvement deservedly finds no advocate among the people."[13] Under these circumstances, the legislators, who had been counting on the surplus of canal tolls to meet a large proportion of the current expenses of the state, were far from willing to reverse the position and place sound fiscal supports under a program of further development.

The new policy of retrenchment was embodied in the "stop law" of 1842 and in the constitution of 1846. The former ordered a cessation of construction on all the canals, except the most necessary repair and maintenance, and imposed a statewide property tax, for the first time since 1827, to replenish the general and canal funds. The constitution provided that the system of public canals, including the two unfinished laterals, must not be disposed of by the state. On the other hand, it placed a heavy burden on their receipts. Charges of $1,850,000 per year, subject to increase under certain conditions, were levied against the canal revenues—for sinking funds for the debts on the original and lateral canals and on the state's railroad loans, and for a contribution of $200,000 to the current expenses of the state—before any of the receipts could be spent for further construction. Moreover, no new debts of more than $1,000,000 were to be incurred except after a popular referendum. Still more decisive was the change in policy with respect to state aid. By unanimous vote the delegates to the constitutional convention adopted a provision prohibiting further state loans to private corporations.

Yet the revulsion against internal improvements was neither complete nor permanent. Local government support of railroads began in 1837 and continued during the depression years. The most important of the early cases were those of the rival cities of Troy and Albany. In the effort to establish itself as the Hudson River terminus for traffic from the West, Troy built the Schenectady and Troy Railroad at a cost of $700,000 and operated it as a municipal enterprise. When this was completed in 1842, the city of Albany countered by pledging $250,000 in aid to the Mohawk and Hudson. Troy's road was well constructed but failed to secure the traffic and was finally sold at a heavy loss. Both roads were eventually absorbed by the New York Central, but the Mohawk and Hudson became the main line. Albany thus won the competitive victory, and it emerged without financial loss both on this commitment and on its earlier loan of $1,000,000 to the

Albany and West Stockbridge for a road to the Massachusetts line. Less successful were the ten cases of local railroad investment in the fifties, which totalled $1,275,000. More than half of this represented another costly venture of the city of Troy, in the stock of the Troy Union; and the others were smaller investments in railroad stock, all but one of which proved financially worthless.[14]

Though new state aid to private enterprise was forbidden by the constitution of 1846, a conditional obligation to the Erie Railroad assumed by legislation of the preceding year involved the state in substantial additional expenditure. Under this act, completion of the road to Dunkirk on Lake Erie in 1851 resulted in cancellation of the original loan of $3,000,000 and assumption by the state of interest payments on the company's other bonds, which came ultimately to more than an additional $3,000,000.

Meanwhile, with the return of prosperity and with the continuing increases in Erie Canal traffic and revenue, investment in the state's public works was resumed both on the enlargement and on the two laterals. For several years the restrictions of the new constitution were partially evaded by the sale of "canal revenue certificates" in anticipation of future tolls; and in 1854 a $9,000,000 loan was approved by popular vote. The Black River lateral was completed in 1855 and the Genesee Valley two years later. A private company built the Junction Canal in the mid-fifties, which increased the usefulness of the Chemung lateral by connecting it with the Pennsylvania canal system. On the other hand the state never finished the Chenango Extension which was intended to perform the same service for the less successful Chenango lateral. The long delay in constructing the Genesee Canal, on which traffic began to decline even before it was completed to Olean, destroyed whatever possibility it might have had of a period of usefulness; and the piecemeal progress on the Erie enlargement, not officially completed until 1862, prevented full exploitation of its steadily increasing opportunity. Yet in spite of these difficulties the operating income on the public works as a whole, which by 1861 had cost the state about $57,000,000, had exceeded the interest on the loans incurred for their construction by nearly $27,500,000.[15]

After more than twenty years of vigorous state action and another twenty years of more restricted state and local activity, New York State was served by three main routes to the West. Of these the Erie Canal remained the most important. The public works also included useful connections to Lakes Ontario and Champlain and a system of branches and laterals which was already seen to be overextended. The other through routes were railroads. Of these the less favorably located, the Erie, had twice been rescued by the state from financial distress. The second was the New York Central, which was mainly the work of private enterprise. It was formed in 1853 by the consolidation of a number of short lines which had joined Albany to Buffalo as early as 1842 and had made the connection to New York City, with a ferry at Albany, when the Hudson River Railroad was completed in 1851. Of the Central's main line only one section received public aid.[16] To these through routes should be added the Delaware and Hudson Canal and a growing network of shorter railroads, which had been initiated by private promoters but were often recipients of public aid.

An analysis of public policy must take into account these varied experiences. Possessing what was potentially the best route to the West, the state of New York acted promptly and decisively, without diversion to side issues, to exploit this major advantage. The Erie Canal, built as a public work at a time when its scope was beyond the reach of private enterprise, was and remained the greatest single triumph of American improvements. The innovation included a sound basis for approaching the European money market;[17] and the establishment of the Canal Commissioners as a permanent agency permitted the development of orderly procedures of construction and administration and at least some progress in devising reasonable estimates, not always heeded by the legislature, of the costs and benefits of new canals.[18] Yet the very success of the Erie Canal tended in certain ways to distort subsequent policy decisions. On the one hand, it encouraged popular demands for building a number of other canals with more dubious economic prospects. On the other hand, the possibility of applying its revenues to general state purposes and thus avoiding taxation was so tempting that legislators were unwilling to provide funds for the completion of improvements at the times at which they would have yielded the greatest return. Success in exploitation made it difficult to think in terms of the economics of development. In the third place, concern for maximizing the tolls from the Erie Canal led the state to restrict the growth of the competing railroads along its route. Little attempt was made to foresee the respective roles of the rival types of improvement or to provide for their coordinated development.[19] When state and local authorities gave assistance to individual railroad companies, they did so on a piecemeal basis without much relation to the creation of a general network of transportation.

A number of other states, as President Madison predicted, found in the Erie Canal "an example . . . worthy of emulating enterprise." [20] They were impressed both by its success as an improvement and by "the revelation that states could borrow in London" for purposes of development.

The example proved particularly stimulating to the neighboring commonwealth of Pennsylvania, which had already a considerable background of planning, agitation, and action in internal improvements. A Society for Promoting the Improvement of Road and Inland Navigation, organized as early as 1789, produced what might be regarded as a Gallatin Plan for Pennsylvania. A series of "memorials" analyzed "the various improvements of which the State of Pennsylvania is susceptible." One was the "Delaware Navigation," which would involve "a happy rivalship between the cities of Philadelphia and New York." Others were the improvements of routes from the Delaware and Susquehanna valleys across New York State to Oswego on Lake Ontario. Most important of all was the western communication with Pittsburgh and Lake Erie. For this last, three possible routes were considered, but the Society's recommendation was an eighteen-mile portage between the Juniata tributary of the Susquehanna and the Kiskiminitas branch of the Allegheny.

For the execution of such a program of improvements, the Society proposed the adoption of "some uniform and consistent plan," with a board of commissioners as its administrative agent and financial procedures varying with the economic circumstances. Wherever possible, "turnpike roads, or toll navigations" should be made by purely private enterprise. On the other hand, the necessary work should be done at the expense of the state where "the smallness of the tolls, the distance from the market and other circumstances may yield the least probability of an adequate encouragement or speedy reimbursement to the adventurers." In intermediate cases, at the discretion of the board, the state might encourage private companies either by loans or by subscription to their stock.[21]

This plan was never adopted as a whole, but assistance was given to a number of individual projects. The state's first appropriation for a turnpike, extended under a law of 1806, was to a company that was building a part of the Pittsburgh Pike. This famous thoroughfare, finished in 1817, competed on equal terms with the Cumberland Road, and for a number of years the two were the nation's principal routes for the shipment of manufactured goods to the West.[22] By 1825, the state was a stockholder to the amount of about $1,800,000 in fifty-six turnpike corporations, as compared with private subscriptions of over $4,-000,000, and Pennsylvania possessed a system of through roads which it considered unsurpassed in the country. The state also held stock of some $180,000 in bridge corporations, as against about $1,600,000 in private subscriptions. On the argument that it would bring the trade of the Susquehanna Valley to the port of Philadelphia, the Legislature authorized a subscription of $100,000 to the Chesapeake and Delaware Canal Company, which received smaller sums from the states of Maryland and Delaware and $300,000 from the federal government. Pennsylvania also held investments of $185,000 in three canal and navigation companies operating within the state. In the Monongahela Navigation, in the western region, the state held the majority of the stock; but its share in the Schuylkill Navigation was about five percent of the whole and in the Union Canal Company about ten percent. By 1825, the latter company had still, after many delays, not quite completed the connection between the Susquehanna and the Delaware Valleys, for which the first surveys had been made sixty-five years before.[23] When New York's canal was opened, Pennsylvania possessed a superior road connection with the West but had not finished even the first stretch of a possible water route across the state.

The challenge of the Erie Canal had a prompt and decisive effect on Pennsylvania policy. Advocates of internal improvements drew two morals from the New York experience: first, that Pennsylvania and particularly the city of Philadelphia would lose the trade of the West if they did not meet the new competition; and second, that a Pennsylvania canal would have similar financial success and free the state from much of the burden of taxation. Few called attention to the great differences in terrain; and some claimed that a Pennsylvania canal, in spite of the mountains to be crossed, would be even more profitable than the Erie because of the shorter distance to the West and a longer season free of ice. In December, 1823, a committee of the legislature cited the approaching completion of the Erie Canal as a consideration making it imperative for Pennsylvania "to bring into active exertion those financial and geographical means with which she is endowed by a bountiful Creator." In 1824, commissioners appointed to explore routes for a canal from the East to Pittsburgh, reported in favor of the Juniata connection. In the same year, still another society was formed, again following old Philadelphia custom, to assume the leadership of the movement.[24] The Pennsylvania Society for the Promotion of Internal Improvements produced an eloquent series of memorials and sent an agent abroad to investigate European canals and railways. Though this indicated an initial willingness to consider the very new alternative of the railroad, the leading spirits in the Society soon came to the conclusion that it was necessary to close ranks in support of a canal, apparently in the belief that complete unity among advocates of a western connection was needed to obtain action while New York's example was still fresh in the public mind.[25] Largely through the efforts of the Society, an internal improvement convention, organized on a voluntary basis but with delegates from almost every county in the state, was held at Harrisburg in August, 1825. This body, by a large but not unanimous vote, declared itself in favor of "opening an entire and complete communication . . . from the Susquehanna to the Allegheny and Ohio, and from the Allegheny to Lake Erie, by the nearest and best practicable route." This, it said, was "indispensably necessary to preserve the character and standing of the State." Finally in February, 1826, a few months after the Erie celebrations, the Pennsylvania legislature adopted an act "to provide for the commencement of a Canal, to be constructed at the expense of the State, and to be styled 'The Pennsylvania Canal.' "

This measure provided for commencement of the work along the Allegheny and Juniata Rivers. As its title indicates, the plan was for a single canal "for the purpose of connecting the eastern and western waters." But an act passed in the following year, though described as providing "for the further extension of the Pennsylvania Canal," in fact reflected a quite different set of pressures and purposes. The "Main Line" of the east-west connection would, if successful, bring great benefits to Philadelphia, to Pittsburgh and to the territory along its route; but other parts of the Commonwealth were no less determined to secure the advantages of public works constructed in their own regions. Though the majority at the Harrisburg convention had urged putting aside "all local objects" in order to permit "an undivided assertion of the public strength," the Main Line advocates found that they could not advance their own undertaking without concessions to win the votes of the supporters of local and regional projects. Accordingly, the law of 1827, while providing for further work on the east-west connection, also authorized the construction of canals on the upper Susquehanna and the Delaware and provided for surveys of nine other projects in the Delaware region and in the upper and lower parts of the Susquehanna Valley.[26] The measure thus resembled New York's great canal law of 1825, with the crucial difference that the latter was passed at the end instead of at the beginning of work on the Erie Canal.

A similar relationship continued throughout the history of the program. By 1835, a Main Line of 359 miles had been completed between Philadelphia and Pittsburgh at a cost of a little over $12,000,000, but nearly $6,500,000 had been spent on the construction of 314 miles of other canals. The Main Line itself consisted of alternate segments of canal and railway. Between Philadelphia and the Susquehanna, use of the recently-completed Union Canal was, on the advice of an officer of the Army Engineers, rejected in favor of a shorter railroad route. For the mountain crossing, once early notions of a great tunnel had been discarded, a canal was recognized to be wholly impracticable. Through travel and traffic therefore required three changes: going from east to west, after 82 miles of railroad, there were 172 miles of canal along the Susquehanna and Juniata, the 36 miles of the Allegheny Portage Railroad, and finally another 105 miles of canal from Johnstown to Pittsburgh. Inclined planes were used at both ends of the Philadelphia and Columbia Railroad, and a battery of ten planes was used to get trains up and down the 2,200-foot summit of the mountain portage. On the two railroads, the state furnished the motive power, but private freight carriers provided their own rolling stock. Thus, if Pennsylvania did not have the single canal of the original plan, it may be said to have operated its hybrid system as much like a canal as it could. Completed at a time when there were barely a thousand miles of railroad in the country, the Main Line was a remarkable engineering achievement but an equally remarkable operational and economic anomaly.

The so-called branch or lateral canals differed among themselves in character and purpose. The Delaware Division, quite unrelated to the others, was built mainly for the transport of anthracite coal, like New York's Delaware and Hudson. Canals on the Susquehanna and its north and west branches connected with each other and with the Main Line. The Beaver and French Creek divisions in the northwestern part of the state, though not yet linked up with the rest of the system, reflected the early intention to provide a through route to Lake Erie.

Further legislation in 1836 and 1838 authorized another wave of "branch line" construction, as well as the replacement of the inclined plane at Columbia on the Main Line. The new undertakings included a North Branch Extension to connect with the canal system of New York, an Erie Extension to complete the route to the Lake, and—still more ambitious—a West Branch Extension intended eventually to provide a second line across the mountains, on one of the alternative routes suggested in the Gallatin Report. Work was also begun on two feeder canals and on the Gettysburg Extension Railroad, which was to make connections in Maryland with the Baltimore and Ohio, and with the Chesapeake and Ohio Canal. Of this work the ironmaster and politician Thaddeus Stevens was the leading advocate and would be—according to his enemies—the principal beneficiary. For these various purposes some $7,000,000 were spent or committed before the program was brought to an abrupt halt, with all the new undertakings uncompleted, by the state's financial crisis in 1842.[27]

In August, 1842, when the state failed to meet its interest payments, a total debt of over $33,000,000 had been incurred for the public works. The financial difficulty was for several reasons more severe than in the case of New York. A major difference lay in the economic and financial results of the improvements themselves. Though the Main Line yielded some operating revenue from the start and though in most years this exceeded the operating deficit of the lateral canals, the net operating revenue of the public works as a whole amounted in 1839 to less than a quarter of the annual interest burden. At the same time the New York canals, carrying well over twice as great a tonnage as those of Pennsylvania, were earning on an investment of comparable size a net return of some $500,000 over current interest payments.[28] There were also differences on the side of financing. Pennsylvania legislators and citizens had been even more reluctant than New Yorkers to resort to taxation for the support of internal improvements. Tax revenues pledged to the canal funds by the act of 1827 were considerably less than in New York State, and an additional tax was not voted until 1840. Moreover, two financial windfalls, coming at a strategic time in the program, encouraged the Pennsylvania legislature to further liberality. The federal Surplus Revenue Act of 1836 allotted nearly $3,000,000 to Pennsylvania. In the same year the Bank of the United States, deprived of its federal charter and function, avoided liquidation by transfer to Pennsylvania jurisdiction. The "price . . . paid for survival" included a bonus of $2,000,000 to the state for improvements and the obligation to grant loans of six or seven million dollars for the same purpose. The Bank, said President Biddle, was to be "made the channel by which capital to any extent may be brought in to aid our Pennsylvania improvements." [29] This relationship, which at first seemed to solve the financial problems of the state's improvements, brought new difficulties with the Bank's successive suspensions and ultimate failure.

Under these circumstances, the incidence of crisis and depression, which in New York subjected the public works to retrenchment and delay, had in Pennsylvania the effect of bringing the program to a virtual close. Ways were found to meet obligations to contractors, and strenuous financial measures restored the solvency of the state[30] and made it less embarrassing for traveling Pennsylvanians to sit down at English tables.[31] But the legislature was much more inclined to dispose of the public works than to appropriate money for them, even in those cases in which a small expenditure would bring into operation an all-but-completed canal. In 1842 the governor was authorized to receive bids for any or all sections of the public works. An act of the following year provided for the disposal without compensation of the Beaver and French Creek divisions to a private company on condition that it complete the connection with Lake Erie. In 1844 the legislature passed an act, confirmed by a closely-contested popular referendum, offering the Main Line for sale at $20,000,000. There were no takers at this price for an operation which at the time was earning less than 3 percent on an original cost of $12,000,000.

With the return of prosperity and increasing transportation needs, state investment for the improvement of the system was resumed. The North Branch Extension was finally completed, the Philadelphia and Columbia Railroad was reconstructed after the replacement of its last inclined plane, the profitable Delaware Division was enlarged, and—less understandably—the Allegheny Portage was replaced by a new railroad, completed some two years after the Pennsylvania Railroad had built its competing line across the mountains. On all these some $6,000,000 were spent between 1846 and 1858. But the policy of disposal had never been abandoned. In 1857 the Main Line was finally sold for $7,500,000 to the Pennsylvania Railroad which promptly abandoned the new Portage Railroad. In the following year the disconnected remnants of the state system were sold to the Sunbury and Erie Railroad Company at a figure finally set, after the railroad had resold the properties, at $3,781,000. Since the state accepted most of its payment in the bonds of the struggling company while the latter was free to collect cash, if it could, from the ultimate purchasers, it is clear that the transaction was planned largely with the purpose of aiding the company to complete its road. The terms on which the two sales were made were hotly contested in the legislature, and opponents attributed their approval to the work of lobbyists and "paid scribblers" for the companies;[32] but it appears that there was little positive sentiment in favor of state retention of the system.

Judged on fiscal grounds, the failure of the public works—whatever their contribution to the general development of the state—was clearly evident. The total investment in their construction had been nearly $33,500,000. Since more than $10,500,000 of this had been for improvements either left unfinished or given away, it appears that the functioning properties were sold for about half their cost. Operating revenues had yielded $8,000,000, but this was far outweighed by the $43,500,000 already paid out in interest. Even disregarding interest to be paid on outstanding obligations, the state's net financial outlay to the time of final sale approached $58,000,000.[33]

Sale of the Main Line to the Pennsylvania Railroad Company in 1857 has been characteristically regarded, by contemporaries and by later writers, as a major triumph of private over public enterprise. Now, said a Harrisburg newspaper, the works will be run by "the best businessmen of the state."[34] Yet more than half of the shares of the corporation which had achieved this triumph were still held by public authorities;[35] and the other purchaser of the state properties, the Sunbury and Erie, was also a product of mixed enterprise. Throughout the period of the state works, public encouragement to improvements in transportation was also being given through stock subscription and occasionally by loans or guarantees to the mixed enterprises. By 1842, state investments in the stock of turnpike companies had risen to over $2,000,000 and in bridge companies to more than $500,000. Even in inland navigation, the principal field of the public works, state holdings in the stock of mixed corporations totalled $850,000, and nearly $400,000 had been subscribed to the stock of four railroad companies. Moreover, the state had found a way of giving aid by indirection; the price exacted for the charter of the Bank of the United States included the requirement that it purchase $675,000 of the stock of certain transportation companies. Meanwhile, local authorities had invested smaller amounts in bridge and turnpike enterprises.[36]

The financial crisis brought about a marked change in the position and policy of the state. In order to raise money to meet pressing obligations, the legislature authorized the sale of the state stock in mixed enterprises. Of the transportation stock, which had yielded very scant return, a little more than $2,000,000 was sold, at not much more than twenty cents on the dollar, and nearly $2,000,000 remained unsold for lack of tolerably satisfactory bids.[37]

When investment in mixed enterprise was resumed with returning prosperity, the movement showed two marked differences. In the first place, the emphasis, as would be expected, was on railroads rather than on other forms of transportation. In the second place, the greater part of the investment was made not by the state legislature, which remained under the shadow of financial difficulties, but by the authorities of counties and municipalities anxious to improve the competitive position of their localities. The state gave no new aid to railroads except through the Sunbury and Erie transaction. On the other hand, local authorities took part in the movement on a scale dwarfing the earlier state participation, and their total subscriptions exceeded $18,000,000. By far the larger part of the total was from the two great urban areas. The city of Philadelphia and urban districts within the county subscribed abut $10,000,000, of which half was for the Pennsylvania Railroad; and Allegheny County and its cities of Pittsburgh and Allegheny subscribed $5,500,000, of which $1,000,000 was for the Pennsylvania. The remainder of the investment of the Pittsburgh area was for other roads radiating from that center. Philadelphia, which made its initial investment in the Pennsylvania when the railroad's nearest point was Harrisburg, continued to support enterprises at a distance that were considered important to the city's trade. It became the principal stockholder in the Sunbury and Erie in northwestern Pennsylvania, and its subscriptions in 1853 to railroads in the northeastern and far southwestern parts of the state were subjects of political controversy and legal contention. The remaining $2,500,000 of local subscriptions was scattered among a dozen railroads by nine counties and ten municipalities.[38]

The subscriptions of the local authorities, together with some public aid from Maryland sources for roads in the Susquehanna Valley, played a major part in the rapid extension of railroads within the state of Pennsylvania. Some fifteen hundred miles were added between 1845 and 1858.[39] Except in the anthracite region, little of this mileage was built by unaided private enterprise. Without local assistance, the Pennsylvania Railroad certainly could not have reached Pittsburgh by the middle fifties in time to compete effectively with the Erie Railroad and the New York Central. The financial results, however, were much less favorable. With the major exception of the Pennsylvania, most of the companies paid small dividends or none at all, and an almost complete cessation of payments during the panic of 1857 led to a partial repudiation of Allegheny County bonds and brought to a head a statewide movement of opposition to mixed enterprise. In that year amendments to the constitution were adopted forbidding state aid to corporations, either by loan or subscription, and prohibiting loans or gifts or stock subscriptions on the part of local authorities. These anti-investment amendments were approved by a nine to one majority in a popular referendum.

Pennsylvania history is thus marked by a complete revulsion against government participation in transportation development. The traditional and almost unquestioned acceptance of the promotional role of the state gave way, as Louis Hartz has shown, to the belief that action in this field was entirely outside the proper sphere of government.[40] With respect to public enterprise, the major turning point occurred during the financial difficulties of the early forties, but the final disposal of the state works coincided almost exactly with the decisive action taken against mixed enterprise in 1857.

The revulsion was partly due to the growing strength of the Pennsylvania Railroad and other corporations under private leadership. However heavily they had depended on public subscription in the past, they now seemed confident that they could raise the funds for further expansion from earnings and private investment. No less important was general disillusionment with the financial results of the public program. For its heavy expenditures on the public works, the state had no assets to show, though most of the works themselves were still in operation under mixed-enterprise or private corporations. Of its subscriptions to mixed enterprise, stocks with a face value of nearly $1,750,000 were still in the treasury but yielded almost no return.[41]

These results were in part the consequences of mistakes in policy and failures in administration. The poor record was to some degree due to corruption and political patronage in the management of the state works,[42] and some of the losses caused by abandonments and forced sales could have been avoided if the necessity of taxation had been recognized at an earlier stage. With respect to the public works, it is often argued that the major mistake was the failure to push through the Main Line to the

West without diversion of effort to less essential undertakings. The early proliferation of branch lines certainly reflected legislative bargains between conflicting local forces rather than any rational determination of priorities. The geographical diversities of interest within the state—among them the Susquehanna Valley's interest in trade downriver to Baltimore as against the Philadelphian's identification of state patriotism with the commerce of his own port—continued to make agreement on a coherent program almost impossible. Yet the financial results of the completed branch lines, taken as a whole and measured either in operating revenues or final disposition, were not much more unfavorable than those of the Main Line; and it is not clear that their contribution to development was greatly inferior. The one highly profitable segment of the state system was the coal-carrying Delaware Division, which might well have been built by private enterprise and would easily have ranked as a "dowager canal."

In retrospect it is easy to see that the Main Line could not possibly have been made an effective competitor of the Erie Canal for the trade of the West. A canal over the mountains was, as a modern writer has said, "a technological monstrosity." No combination of linked canals and railroads, which had to surmount an elevation of more than two thousand feet, could match the costs of a waterway which at no point had to be raised above the level of Lake Erie.[43] It is perhaps ironic that the planners of internal improvements in Pennsylvania, who were somewhat more receptive than those of New York to the innovation of the railroad, made use of it in this limited and ineffective way at a time when its potentialities could not be clearly seen. Pennsylvania's combination of old and new, "elucidating the defects peculiar to both methods of transit,"[44] could not solve the problem. The state's disadvantages in terrain could not be substantially overcome, nor could its advantages in distance to the West be effectively utilized until it became practicable to build an all-rail route. This was not accomplished until more than twenty years after the completion of the Main Line, and then by mixed rather than public enterprise.

In the Pennsylvania case, moreover, effective leadership under mixed enterprise came almost entirely from the private side. In part this was a matter of clear legislative intent. An early act governing the state's bridge and turnpike subscriptions provided that, in cases in which the state held half or more of the stock, it should appoint two of the five directors; and a measure relating to the Pennsylvania Railroad subscriptions, the greatest single case of local action, contained the express provision that "a majority of the board of directors shall, at all times be elected by the private stockholders." [45] Within this general framework there were variations in the degree to which the public authorities attempted to participate in the management of the enterprises which they supported. In a number of cases the state made no provision for the appointment of government directors, and there were instances in which state directors were refused admittance to meetings of the boards. The principal concerns of the less powerful local authorities were expressed in provisions in the acts requiring that the railroad should "pass through the said borough" or that the funds "be expended in the said country." [46]

In Philadelphia, where the initial subscription to the Pennsylvania Railroad was authorized after an election contested by "Railroad" and "Anti-Railroad" slates of candidates, the City Councils made active attempts to watch over their interests in this and other roads. They withheld their promised subscription to the Sunbury and Erie until an unacceptable board of managers was replaced and expressed vigorous opinions on the question of the Pennsylvania Railroad's investments in the securities of connecting railways in the West. In the 1853 stockholders' meeting, the representatives of Allegheny County opposed purchase of stock in the Marietta and Cincinnati; but those of Philadelphia supported it, acting under a resolution of the City Councils declaring that the measure was essential to secure to their city "her just proportion of the trade of the West and South-West" which was endangered by "the great efforts now making by the cities of New York and Baltimore." [47] Even in the revulsion year of 1857, an attempt was made at the time of sale of the public works to assure perpetual representation of the city and Allegheny County authorities on the board of the Pennsylvania Railroad on the ground that "so important a work should not be managed" wholly in the interests of its private stockholders.[48] But what was at issue was minority representation—in this case five out of fourteen directors—and initiative and effective control had rested from the beginning in private hands.

Pennsylvania's early improvement Society, it will be recalled, recommended that the work should be carried forward under a "uniform and consistent plan," supervised by a board of improvement, and by a judicious division of labor between private, public, and mixed enterprise. Down to 1857, the three methods continued to be employed, though in varying proportions for the different means of transport. The canals were mainly built by the state and the railroad network was supplied partly by unaided private enterprise and partly by the state but more largely by mixed enterprise resting heavily on the subscriptions of local authorities. On the other hand, the first parts of the Society's recommendations went largely unheeded. The board of improvement was never established. Little consistency of planning was achieved within the program of state works and still less between it and the other measures. Pennsylvania's vigorous but often conflicting efforts toward internal improvement went forward with little of the coordination advocated by its eighteenth-century memorialists.

Maryland's way to the West was by the Potomac and then across the mountains to the Monongahela and the Ohio. This was the route of Braddock's Road and of George Washington's greatest interest, and it followed the shortest of Gallatin's lines from the Atlantic ports to the western waters. Two great enterprises attempted to reach this goal: the Chesapeake and Ohio Canal and the Baltimore and Ohio Railroad. For each the first earth was turned in competing ceremonies on the Fourth of July, 1828, and their rivalry was later dramatized in a legal contest for the use of a narrow part of the Potomac Valley. Though one represented the older, and the other the newer technology, both were the products of mixed enterprise, with the canal receiving its main support from the state and the railroad from the city of Baltimore. In part because of this rivalry,[49] Maryland never followed New York and Philadelphia in employing the method of public works; and aid to these and to certain other enterprises continued to be given by state and municipal subscriptions and loans and guarantees.

When work began in 1828 on the two major projects, one of them had behind it a long period of agitation and endeavor. The Chesapeake and Ohio Canal was the successor of the old Potomac Company, of which George Washington was the first president. Both Maryland and Virginia had supported the enterprise, with Maryland's investment amounting to about $150,000. The company improved navigation on the river from Cumberland down, with some work on the tributaries, and by the time of Gallatin's Report had built canals around the five major falls on the Potomac. But it did not attain its modest charter objective of providing a safe channel in all seasons for vessels carrying fifty barrels of flour. Twenty years later little more had been done and the company was bankrupt.

The new and more ambitious project of a completely independent canal that would extend up the valley to Cumberland and eventually across the mountains to Pittsburgh, was launched in the eighteen-twenties by a combination of federal, state, municipal, and private effort. Early interest was strong in Virginia but the terms for a state subscription offered in 1823 were never met. On the other hand, Maryland refused participation in that year, largely because of opposition from Baltimore, but subscribed $500,000 in 1825 on condition that the charter provide for a branch canal to Baltimore. The decisive actions were the federal government's subscription of $1,000,000 (its largest financial participation in the mixed enterprise of the time) and the permission it gave to the District cities for municipal subscriptions of $1,000,000 by Washington and $250,000 each by Georgetown and Alexandria. Private individuals subscribed only a little more than $600,000 as compared to the $3,000,000 of public money. The federal government also provided an engineering survey of the proposed route, the first carried out under the General Survey Act, and its leading role in the initiation of the project was symbolized when the first shovelful of dirt was turned by President John Quincy Adams.[50]

The initial impetus and leadership were, however, not long maintained. The company decided, in part because of the recommendation of the Army Engineers, that the canal should be built to higher specifications in width and depth than the other canals of the day. For this and other reasons it soon became evident that the costs, even to Cumberland, would exceed the original subscriptions. The latter proved partly uncollectible; nearly a third of the private subscription was never paid, and the obligations of the District cities had finally to be assumed by the federal government. Labor was unexpectedly difficult to obtain, and the problem was only partly solved by recruitment from abroad. In 1832 an epidemic of cholera swept through the workers' shanties. The legal contest with the Baltimore and Ohio Railroad for control of the narrow valley above the Point of Rocks, where there was barely room to build both canal and railway, caused serious delays in construction even though the Chesapeake and Ohio's prior claim was finally sustained by the courts in 1832. At the end of that year, the company was out of funds and construction was halted far short of the hundred miles which, according to the terms of the charter, were to be completed by 1833.

Appeals for further aid from the national government were fruitless. Jacksonian opposition to federal improvements had taken the place of Adams's advocacy, and the effect of the change was to leave the Chesapeake and Ohio canal project "stranded on the southern shore of Maryland." [51] In the emergency it was the state that took over the responsibility. An act of 1833 forced upon the rival companies a compromise providing for joint construction of canal and railway between Point of Rocks and Harper's Ferry. In 1834 the state made an additional subscription of $125,000 and in the following year a loan of $2,000,000. An omnibus bill of 1836, which also provided support for the Baltimore and Ohio and other undertakings, made the state the majority stockholder in the Chesapeake and Ohio with an additional subscription of $3,000,000. Another measure in 1839 brought the state's total subscription to $5,000,000. "The great national project" had become "a Maryland work." [52]

With this support, work on the canal was resumed. Engineering difficulties increased, but the Board continued to maintain its policy of expensive construction and high technical standards, insisting the "false notions of economy" were inappropriate in a work designed not "to subserve the purposes of the present day or century, but . . . to endure for all time." [53] Labor problems were intensified by the outbreak of pitched battles between the laborers from the two Irish counties of Cork and Longford. Moreover, the company's apparent assets could not always be realized. Maryland state bonds, like those of the District cities, were sometimes unsalable or could be disposed of only at heavy discounts; and the company's final realization from the $7,000,000 of state aid was only a little more than $6,000,000. At the various periods of financial stringency, the company employed unorthodox methods to keep the work going. In 1834, from 1837 to 1838, and again from 1839 to 1841, the company met part of its obligations, particularly to its contractors, in its own notes or "scrip," each time issued in larger amounts and reaching finally a total of over $1,400,000. By these and other means work was maintained through periods of depression, but finally came to a halt in 1842. One hundred and thirty-five miles had been completed; but the canal's first objective—Cumberland—was still fifty miles away.

The rival enterprise had by this time already reached Cumberland. The Baltimore and Ohio Railroad, organized in 1827 under the stimulus of the Erie Canal, represented the effort of the merchants and the city of Baltimore to retain or regain their position in the commerce of the West. In this it resembled Philadelphia's parallel efforts to meet the competition by means of the State Works. Its merchant promoters, however, unlike the Pennsylvania legislators, made the bold choice of the newer technology. Rejecting the alternative of a branch canal to the Chesapeake and Ohio, they decided to attempt to build a railroad over the mountains, at a time when England's short lines were still in an experimental stage.[54] The original subscription consisted of $1,500,000 from private investors and $500,000 from the city of Baltimore; and in the following year the capitalization was doubled with an additional $1,500,000 from individuals and $500,000 from the state of Maryland. Requests for aid from the federal government were refused, but the state added $500,000 in 1833 for a branch to Washington.

As a pioneering enterprise, which Henry Varnum Poor once called "the *Railroad University* of the United States,"[55] the Baltimore and Ohio had many problems of construction and operation to solve. Yet once it had learned a satisfactory application of steam power, after early use of horses and even an experiment with sail, construction went forward more rapidly and at lower cost than in the case of the canal. A first short section was opened for traffic in 1831, and in the following year the road reached the Point of Rocks, some seventy miles from Baltimore. After settlement of the controversy with the Canal Company, the line was extended in 1834 to a point across the river from Harper's Ferry.

Here, however, there was a check to further progress. The canal interests were anxious to keep the railroad out of competition along the river to Cumberland; and the railroad itself might, given sufficient encouragement from Virginia, have chosen to take an alternate route to the West by way of the Shenandoah Valley. Private funds to finance the next stage of construction were not forthcoming. Under these circumstances, the company again appealed to state and city. Each responded in 1836, the peak year of public appropriations, with a subscription of $3,000,000. As a result of these actions, an enterprise that had begun three parts private and one part public now had a public majority of stock and directors. But bonds authorized in 1836 could not be sold under the crisis conditions of 1837 and 1839; and the railroad company, like the canal, began in 1839 to anticipate receipts by circulating its own currency. Nearly $1,500,000 of its "railroad notes" were issued by September, 1841, and the city of Baltimore accepted them in payment for all city dues.[56] By these expedients, work on the road was continued without interruption, and it reached Cumberland, 178 miles from Baltimore, in 1842.

On railroad and canal, work was halted in 1842. There was a long hiatus in construction and in public aid to these and other improvements. In the case of the state government, which defaulted on its bond interest in 1842 and was unable to resume full payment until 1848, the revulsion was permanent. No new financial aid was given to any company; and the members of the Reform Convention of 1851, commenting bitterly on those who had thought that investments in internal improvements would make the state a taxless "El Dorado," adopted constitutional amendments prohibiting state works and loans or gifts to improvement companies.[57] In the case of the city, which remained solvent, no further appropriations for improvement were made until 1852.

In spite of these circumstances, work was finally resumed on the canal. After another failure to obtain federal aid, the company managed in 1845 to secure from the Maryland legislature—by the margin of a single vote in the face of bitter opposition from the representatives of Baltimore—an act waiving the state's prior lien on the canal in order to permit the company to issue its own first mortgage bonds. Guarantees of $300,000 from Virginia, $25,000 from Georgetown, and $50,000 from the city of Washington were secured for new construction, with an additional $200,000 guarantee from Virginia for necessary repairs. With this aid, the canal was finally completed to Cumberland in 1850 at a total construction cost of almost $10,000,000, of which the state of Maryland had supplied a little more than three fifths. The usual celebrations were held, but the triumph was a limited one. Standards of construction had been maintained in spite of engineering difficulties, and the canal was built to the most generous dimensions of any in the United States. But the original purpose of a waterway to the Ohio had been tacitly abandoned. The canal reached only the foot of the mountains, and that at a time when the railroad had already taken much of its prospective trade. Canal tonnage rose from about 100,000 in 1850 to about 350,000 at the end of the decade, and tolls from $65,000 to over $180,000; but net receipts yielded no surplus over interest, and by the outbreak of the Civil War the company was close to bankruptcy.[58]

By contrast, the railroad succeeded in crossing the mountains without further aid from the state or from the city of Baltimore. The city of Wheeling, its Ohio River terminus, subscribed $500,000. A substantial part of the remaining funds came from the ploughing back of the Baltimore and Ohio's own operating revenues, since the company in this period issued its dividends mainly in bonds and stock. Another part came from the disposal between 1849 and 1851 of the predepression Maryland bonds, which had finally become negotiable with the return of prosperity and the restoration of the state's credit. The remainder came from the company's own borrowing, done largely through Baring Brothers, which regarded the Baltimore and Ohio as "the best railroad investment in the United States." [59] The construction cost from Cumberland to Wheeling amounted to about $6,600,000 as compared with some $7,600,000 from Baltimore to Cumberland. Of this total a little more than half had been obtained from government sources.[60] The line was completed in 1853, a little after the New York and Erie and a little before the Pennsylvania Railroad; and Baltimore had achieved its connection with the western waters.

This was by no means the end of Baltimore's aid to the B. & O. system. In 1852 the city endorsed $1,500,000 of the bonds of the Northwestern Virginia, which was to be the B. & O.'s alternative route to the Ohio; and additional loans were made in 1859, paid for out of dividends on the city's B. & O. stock, of $135,000 to the Northwestern Virginia and of $20,000 to the Union Railroad across the river in Ohio, which formed part of the connection to Cincinnati. The parent company itself received a loan of $5,000,000 in 1853 for the rehabilitation of the road. In 1856 a loan of $1,000,000 was made to the Pittsburgh and Connellsville, which had already received aid from Allegheny County and from several Pennsylvania municipalities and which would ultimately provide the system's connection between Cumberland and Pittsburgh.[61] This road was far from completion at the outbreak of war, but by this time the B. & O.'s main line to Wheeling and the more southerly route to Parkersburg and Cincinnati were both effective lines of transportation to the West.

Neither state nor local aid was confined to these efforts to reach the western waters. As early as 1800 the state loaned $30,000 to the Susquehanna Canal Company; a decade later it subscribed $15,000 to the stock of two turnpikes; and in the twenties it made a $50,000 subscription to the Chesapeake and

Delaware Canal.[62] A much larger amount of assistance was given to the Baltimore and Susquehanna Railroad which was intended to maintain Baltimore's position as the natural outlet of Pennsylvania's rich Susquehanna Valley. After a stock subscription of $100,000 in 1830, Maryland in its major period of improvement legislation, 1835 to 1839, added three loans bringing its total aid to about $2,250,000. A canal by-passing the obstructions of the lower river, the Susquehanna and Tidewater, received a $750,000 loan in 1839 and was completed in the following year.

The state's other appropriations were for more questionable undertakings. The Eight Million Dollar Act, defeated by a single vote in 1835 and passed after great controversy in the following year, provided major sums for the two routes to the West but added to them appropriations of $500,000 for a cross-cut canal to connect Baltimore with the Chesapeake and Ohio, the same amount for a canal from Annapolis to the Potomac, and $1,000,000 for the Eastern Shore Railroad. The two canals were never started, but $300,000 of the second appropriation was transferred to the Annapolis and Elkridge Railroad, which also received aid from the city of Annapolis; and the state later guaranteed the interest on $80,000 of the company's bonds. The Eastern Shore Railroad, projected in a little-developed area, received almost no private subscriptions; but the state paid some $150,000 toward the beginnings of construction and in payment of the debts of the soon-bankrupt company. An act of 1848 aided the Baltimore and Susquehanna by granting a moratorium on its arrears of interest, and a measure of 1860 revived the balance of the old subscription to the Eastern Shore, though with little immediate expenditure. But new financial aid to internal improvements had ended with the state's financial difficulties and with the constitutional provision of 1851.

The remainder of Baltimore's aid was almost entirely devoted to the Susquehanna route. A subscription of $380,000 was made in 1837 to the Susquehanna and Tidewater Canal. Larger amounts went to the railroad. After matching the state's subscription to the Baltimore and Susquehanna in 1830, the city added in 1837 and 1838 loans totaling $850,000 and in the following year gave the company $250,000 in its own bonds in exchange for the same amount in unsalable bonds of the state of Maryland. In 1848 it relinquished its claim to arrears of interest from the company, and in 1855 endorsed $500,000 of the bonds of the B. & S.'s northern extension, the York and Cumberland. Construction of the system on Pennsylvania soil required a series of conflicts in the legislature at Harrisburg with the representatives of Philadelphia, which the road's friends called the "City-of-Want-all-Trade";[63] but the railroad reached the main line of the Pennsylvania State Works in 1840 and, after reorganization as the Northern Central, was completed to Sunbury on the upper river in 1858. In 1860 the City Councils gave their first aid to a new enterprise, the Western Maryland Railroad, authorizing a bond endorsement of $500,000, of which $200,000 was executed.

Between the state and the city programs of internal improvement, there were marked contrasts in planning, in continuity, and in results. As heir to the old dream of a canal from the Potomac to the Ohio, which the Army Engineers had so recently pronounced practicable, Maryland perhaps could not have been expected to foresee the canal's obsolescence, although the railroad had already reached the river by the time the state assumed major responsibility for the Chesapeake and Ohio. It is, however, hard to understand the insistence on "timeless" construction under these circumstances. Prompter completion on a less elaborate scale would have given the canal a longer period of usefulness in the coal and other trades. Moreover, an administration less subject to the political spoils system might well have shown somewhat better operating results in the fifties. The division of the state's aid between the two rival enterprises had some basis in the belief that they had different functions to perform (the canal to carry the heavy products of the farms and mines and the railroad to carry passengers and lighter and more perishable goods).[64] The actual decisions, however, seem to have reflected the compromise of conflicting interests rather more than any rational assessment of advantages.

Sectional influence was even more conspicuous in determining the allocation of the other parts of the state's program of assistance. The discussions and decisions of 1836 provide an extreme example. The House Ways and Means Committee presented a report stating the case for "a general system of improvements [of which] the principal axis" would consist of the routes to the West and up the Susquehanna,[65] but the bill reported by the Committee and adopted by the legislature appropriated nothing for the Susquehanna route and appealed instead to the votes and interests of other areas by including the two obsolete canal projects and the premature railroad for the Eastern Shore. Of a total of some $16,500,000 of aid authorized by the legislature, nearly $1,500,000 remained unextended because the projects themselves were abandoned. Moreover, the legislature's false hopes of quick returns and its failure to back its appropriations by an adequate system of taxation and revenue reduced the effectiveness of the aid given and made it impossible for the state to continue its support for the enterprises which it had wished to encourage.

By contrast, the record of the city's aid is one of clearer purpose and more sustained effort. Without the necessity of composing sectional differences, the City Councils were able to follow a policy directed toward securing and maintaining the commercial interests of the city and of its ambitious merchants. Early advocates of the Susquehanna railroad proclaimed their determination to make Baltimore "the Great Central City of the Union."[66] The Joint Committee on Internal Improvements of the City Council, in an 1836 report which laid down the general principles and the geographic directions of the program, defined the issue in the following terms: "whether the city of Baltimore is to be reduced to a place of comparative insignificance in a commercial point of view, or to assume a position, equal if not superior to that of any other city in the Union." The Joint Committee remained the Council's agency for investigation and recommendations on improvement projects, and the city was represented on the B. & O. and B. & S. boards by directors who took an active part in the proceedings, and made annual reports. In 1860 the city joined with the state in resisting the effort of the company's president to reduce the public representation to a minority of the board.

To this improvement effort the city committed large resources. Total authorizations for stock subscriptions, loans, and endorsements, amounted to more than $13,000,000. Service of the debt incurred for these purposes required a heavy current outlay, for which a special internal improvement tax was levied from 1839 on. Net payments on improvement account, after subtracting dividends and interest received, averaged more than a quarter of entire municipal expenditures from 1843 through 1850, rising in the last of these years to nearly a third, and then declining as the B. & O. resumed payment of cash dividends. Very little income was received from the other companies, and the returns on the Northern Central were particularly disappointing; but by 1861 there was every reason for the city to regard its ownership of Baltimore and Ohio stock as a good financial investment as well as a bulwark of its commercial position.[67]

For a city much smaller than New York or Philadelphia, it was an extraordinarily ambitious undertaking to build a two-pronged railroad to the Ohio, so largely with municipal resources, and to venture on the construction of a third prong. This was an era in which the rivalry of competing cities was a familiar element in the promotion of American improvements; but it should be noted that Baltimore planned its program with unusual consistency of purpose, conducted it with fiscal responsibility, and during this period devoted to it larger funds than any other city in the nation.

Internal improvements in Virginia presented a quite different pattern. In this case a carefully contrived statewide system of mixed enterprise competed for attention and support with the drive for a through route to the western waters. The Potomac lay on Virginia's northern border; and the third of Gallatin's pairs of rivers, the James and Kanawha, offered another route to the West entirely within the state. As early as 1784, George Washington urged development of communication with the West "by one or both of the rivers of this state, which have their sources in the Appalachian mountains." [68] His own particular interest was in the Potomac, but a much greater part of Virginia's effort was to be expended on the James. Yet the state's improvement program as officially adopted provided for the extension of aid on equal terms to enterprises in all parts of the Commonwealth. As a result the record shows a series of compromises between the principle of even-handed assistance and the necessity for large outlays if the James-Kanawha connection was to be made.

Each of the two tendencies was embodied in a notable early report to the Virginia legislature. The first of these was presented in 1812 by a commission to examine the James-Kanawha route, of which the chairman was Chief Justice John Marshall. On part of the way there was already a company in operation. In the twenty years following its incorporation in 1785, the James River Company had built canals around the falls at Richmond to connect with tidewater and had improved the navigation of the river for some 220 miles upstream. On a capitalization of $140,000, of which the state had subscribed one half, the company—after a number of lean years—was by this time paying annual dividends of about twelve percent. Contentedly collecting tolls on the existing works, the company had little incentive to meet the growing demands for further improvements on the river; and it possessed neither the funds nor the authority nor the ambition to extend the route across the mountains. Appointment of the commission reflected the legislature's concern with this larger aim.

John Marshall and his fellow "Commissioners to view certain rivers within the Commonwealth" proceeded—evidently during recess of the Supreme Court—to examine the upper part of the James River, a route across the mountains, and then the Kanawha tributaries, the Greenbrier and the New, as far as the Great Falls of the Kanawha. They concluded that no extraordinary difficulties would be encountered in the improvement of navigation on the upper James and the Greenbrier or in the construction of a twenty-eight-mile turnpike over the mountains between them. They had more doubts about upstream navigation on the New River, because of the velocity of the current, but on this took courage from the reflection that the steamboat, then in its infancy, would doubtless be made more efficient and powerful.

As to the value of the work, the commission's findings were emphatic and positive. Opening the communication would be a great stimulus to agriculture on the immediate route. Still more important was the objective of winning the trade of the Ohio Valley. Warning the legislature of the possible effect of "the immense works meditated in New York," the Commissioners argued that this trade would not use the roundabout lines of communication through New York State or by the mouth of the Mississippi if Virginia seized its opportunity to open up the much shorter route by the James or the Potomac:

In a contest for this interesting prize, the states through which these rivers run have geographical advantages, the benefit of which they can lose only by supineness in themselves, or by extraordinary exertions in others.

To this appeal to the state's competitive spirit, the Chief Justice and his colleagues added that completion of this "great central channel of communication by water" would be of value to the entire nation and would serve to "cement more closely the union of the eastern and western states." [69]

The second of these early reports was presented in 1815 by the assembly's Committee on Roads and Navigation under the chairmanship of Charles F. Mercer, who was later a congressman and president of the Chesapeake and Ohio Canal Company. The report's concern was with the extension and systematization of the movement for internal improvements, which already had behind it a body of tradition and experience. In colonial Virginia the legislature had from time to time authorized associations of gentlemen to raise subscriptions for the clearing of rivers or the building of the roads, sometimes adding the privilege of a lottery. Thomas Jefferson recalled with satisfaction his efforts in initiating an improvement in the Rivanna River.[70] In 1785 the state of Virginia had taken part in launching the Potomac as well as the James projects, and it purchased 120 shares of the stock of the Potomac Company. It had also made subscriptions to the stock of a local river-improvement company and of a turnpike; and it had done the same for the Dismal Swamp Canal Company which was to make a connection with the waters of North Carolina.

With this background, the Committee felt that it could take for granted the general acceptance of governmental responsibility for internal improvements. Though there might have been differences of opinion on "the expediency of controlling the voluntary direction of the wealth and labor of individuals . . . there never," it said, had "existed a doubt, but that it is the duty, as well as the interest of every good government to facilitate the necessary communication between its citizens." Next perhaps "to the enjoyment of civil liberty itself" were the blessings which governments could secure to their people "by good roads, navigable rivers and canals." The argument applied with particular force to Virginia, which had "so many advantages to improve and so many obstacles to overcome" and which had allowed so much of her commerce to "pass out of her hands to enrich the coffers of her neighbors."

All this might equally have appeared in the Marshall report. What is distinctive in the second document is its concern with improving the method of state action:

Should the General Assembly determine to patronize by application of the public revenue all such works as are likely to be of great public utility, it becomes important to decide whether an improvement may not be made in the mode heretofore pursued, of extending to them that patronage.

Two difficulties had to be met: blunders in the operation of particular enterprises, and mistakes or favoritism in the choice of projects to be undertaken. The Committee believed that all the canal companies thus far organized in Virginia had suffered "for want of skill in their conduct." Their directors had generally been "public-spirited private gentlemen" serving without compensation. No company had employed or would be able to employ the full-time service of a trained engineer. This was a want the state must supply.

The second problem was that of the choice of projects for support. This was difficult for the legislature, both because of lack of expertness and because of "the variety of opinions always existing in an Assembly representing many local interests." To meet this the Committee made two recommendations: a board should be established to "collect and prepare for the General Assembly, the facts and information necessary to cast" sufficient light upon each application for support; the support should be given not by special appropriation but from a general improvement fund set up in advance. Though each subscription would still require a specific act of the legislature, it should be made under uniform conditions applying to all approved projects.

In formulating these conditions the Committee made explicit its philosophy of mixed enterprise. It could never be to the state's interest "to become the sole proprietor of all the public works within its territory." There might be cases in which state ownership of a particular road or canal would be advisable "in order to accomplish some great commercial or political purpose"; but this should not be the general rule, since experience shows that internal improvements "will be more economically made, and better repaired, if their management be left to the individuals who subscribe to their stock with a view to private gain."

What, then, should be the functions of the state? Since there are many works essential to its prosperity which cannot secure sufficient capital and which cannot be expected to yield returns until population and commerce follow the line of the improvement, the state must provide financial support and it must also exercise a degree of supervision. It should, however, perform both these functions in such a way as to give the greatest possible encouragement to private and local initiative. Its control over the mixed enterprises "should extend no farther, than to correct or prevent such abuses on the community at large, as might be apprehended from the too eager incentive of gain." Moreover, the terms of its financial contributions should be devised to give them the maximum effectiveness in stimulating private investment: "The Commonwealth should subscribe so much, to their stock, and on such terms as will suffice to elicit individual wealth for public improvement."[71]

It was the second of these two reports which resulted in the earlier legislative action. The recommendations of the Mercer Committee were fully embodied in the act of February 5, 1816, creating the Fund for Internal Improvements and the Board of Public Works. The Fund was to include the state's small existing investment in improvement companies and its much larger holdings in bank stock, together with any future bonuses for the extension, modification, or creation of bank charters. The initial assets, taken at face value, were as follows:

7,947	shares	Bank of Virginia	$794,700.00
3,334	"	Farmers Bank of Virginia	333,400.00
250	"	James River Co.	50,000.00
125	"	Appomattox Co.	12,500.00
70	"	Dismal Swamp Canal Co.	17,500.00
70	"	Potowmac Co.	31,111.11
125½	"	Little River Turnpike Co.	12,550.00
			$1,251,761.11

Income from these securities, and any returns from new enterprises supported by the Board, were to constitute a permanent revolving fund to be used for subscription to the stock of improvement companies.

This fund was to be managed by a Board of Public Works consisting of the Governor as president *ex officio*, the Treasurer, the Attorney General, and ten private citizens selected to give representation to each of the state's principal geographic areas. The Board was to employ a Principal Engineer and with his aid was to examine the proposals for aid to particular improvements. Its reports must include "the surveys, plans and estimated expenses of such new works as they may recommend to the patronage of the General Assembly."

For such patronage the act laid down specific terms. The sponsors of each project approved by the legislature must raise by subscription three fifths of the amount needed. The Board would then subscribe the other two fifths and appoint its proportional share of the directors. Dividends on the state's portion of the stock, however, would be deferred until the private subscribers had received six percent on their investment. By this device the Committee believed that "a much smaller subscription of public money [would] suffice to draw forth private enterprise." The purpose of the measure was to give systematic encouragement to individual improvements. Initiative was deliberately decentralized, and the essence of the system was "to call on each section of the State for their three-fifths, who apply for their portion of the public funds." [72]

The recommendations of the Marshall Commission were translated into legislation much less promptly. Plans for the prosecution of the James-Kanawha enterprise were not "mature for the decision of the Legislature" until early in 1820.[73] By this time, after recommendations by the Board of Public Works and surveys by its engineers, the proposals included the replacement of part of the existing improvements by an independent canal, the construction of a canal along the upper James to the foot of the Alleghenies at what is now Covington, the building of a turnpike to the Great Falls of the Kanawha, and the improvement of the lower river to its junction with the Ohio. "Improvements of such magnitude," said the Board, could "properly be undertaken by the General Assembly alone." [74] The state assumed the full responsibility—in 1820, by making the James River Company its agent, and in 1823, by completely taking over the management of the company. Stockholders in the old company were guaranteed twelve percent dividends for the first twelve years and fifteen percent thereafter. On these improvements the state borrowed and spent about $1,300,000. With this sum, a canal of some thirty miles was built from Richmond to Maiden's Adventure, superseding earlier works, and another canal of seven miles was built at the point where the river breaks through the Blue Ridge. Some work was done on the lower part of the Kanawha River. The Kanawha Turnpike was built across the mountains and, at the end of the decade, extended to the Ohio River with a total length of two hundred miles. This became a well-traveled highway, and the canals on the James made possible some increase in traffic and decrease in tolls. But this fell far short of the aim of a canal to the foot of the mountains; and discouragement over meagre results and high costs, together with opposition from other sections of the state, led before the end of the eighteen-twenties to an almost complete stoppage of appropriations and of new construction.

Meanwhile the Fund and Board had begun to operate according to the original plan. Subscriptions from the revolving fund were made to a number of undertakings, and holdings in improvement stocks rose to some $800,000 in 1831.[75] But progress was neither as rapid nor as widely diffused as the authors of the plan had hoped. For this there were several reasons. As the Board pointed out in its second report, "the usual contributions from individuals" could hardly be expected from the sparsely settled valleys of the Kanawha and the Ohio. Returns on the new investments were very small, though income from the banks tended to increase. From 1825 on, moreover, interest on the state's loans for the James River project was charged against the Fund; and the deficits incurred on this account amounted to over $200,000 by 1831. The James River debt, said Governor Tyler "operates as a moth on the fund set apart for the beneficent object of State improvement, in violation of the principle on which the fund was created"; and the Board itself pointed out that its income "was totally insufficient to effect any extended scheme of improvement commensurate with the great interests of the State." [76]

Under these circumstances two significant alterations were made in state policy. The first was to permit the Board to borrow money on the credit of the state for its subscriptions to improvement corporations—a power granted first in 1832 for a small loan and later extended and made general by legislation at the end of the decade. The second was to build certain improvements, particularly roads beyond the mountains, entirely with public funds; modest investments in "state works" began to appear in the accounts in 1837.[77]

Activity on the James River route was revived through another reorganization. Since the legislature was unwilling to appropriate funds for continuation of the enterprise as a state work, the friends of the project secured the organization in 1835 of a new mixed enterprise, the James River and Kanawha Company. Its capitalization of $5,000,000 included the property of the state-owned James River Company, valued at $1,000,000, and $2,000,000 in new funds from the state. Private subscriptions were very hard to obtain, in spite of eloquent appeals by Chief Justice Marshall and others, and in spite of the agreement that the state was never to cast more than a minority of the votes in stockholders' meetings. The remaining $2,000,000, as finally secured, included a subscription of $650,000 by the city of Richmond, obtained after close struggles both in the city and in the legislature; a subscription of $100,000 from the city of Lynchburg; and $500,000 subscribed by the Bank of Virginia under pressure from the state.

The new company completed a continuous improvement along the James from Richmond to Buchanan, a distance of 197 miles, making little use of the old works and relying on slackwater navigation for only a small part of the route. The canal was opened to Lynchburg in 1840 but did not reach Buchanan, fifty miles further on, until eleven years later. The company also rebuilt the Richmond Dock into an effective outlet to tidewater, constructed certain side connections, and did some work on the Kanawha. It began but never completed the section of the canal that was to take it to the foot of the mountains at Covington. Throughout the period the company was in continual financial trouble, in part because of the difficulty of disposing of the municipal and state bonds during depressions, and in part because of the unexpected costliness of the work. By 1860 it had borrowed some $5,500,000 mainly from the state or with the state's guarantee; and its net operating income, though substantial in the fifties, had fallen short of interest payments by about $500,000. The canal provided an outlet for the substantial agricultural production of the Great Valley, and its tonnage reached a peak in 1860, with an amount more than four times that of the state's largest railroad. By this time, however, its net income had already begun to decline as a result of railroad competition.

The charter of the James River and Kanawha Company made provision for the possibilities of a railroad from the head of the canal across the mountains or of a continuous railroad line from Richmond to the Ohio River. Even from the beginning there was some sentiment in favor of the latter alternative; but the management, though sometimes willing to concede the utility of a railroad as a "mere portage," maintained its belief in the ultimate construction of an all-water connection. As late as 1860, plans for continuing the canal to the Kanawha, either by a further reorganization of the company or by its sale to a firm of French engineering contractors, received the approval of the legislature and the company.[78]

The remainder of the state's aid went almost entirely to railroads. The first major enthusiasm for railway building coincided with the decision to borrow heavily in the late thirties, and the railroad companies shared the proceeds of the loans with the James River and Kanawha in what the latter's historian describes as a log-rolling operation.[79] New investment averaged about a million dollars a year from 1836 through 1839. In addition to stock subscription, the state sometimes gave further encouragement by loans to the companies or guarantees of their bonds. In the depression years of the early forties, these loans were converted into preferred stock; and the amount of new public aid, as in other states, dropped almost to nothing by the middle of the decade.

In sharp contrast with Maryland and Pennsylvania, however, Virginia increased its volume of state aid at the end of the eighteen forties; and the terms of subscription were made even more generous than before, with the state contributing three fifths of the capital. In 1851, the Board boasted that Virginia was at last "awake to the transcendent importance" of internal improvements and that the commonwealth stood "not prostrate in repudiation, after the example of so many states . . . but with her bonds commanding high premiums in the markets at home and abroad."[80] For the years 1850 to 1861 inclusive, the Board's borrowings averaged more than $2,000,000 a year, with peaks of almost $4,000,000 in 1853 and 1854. From 1837 on, moreover, the state had made a practice of meeting the Board's deficits on current income account from appropriations; and after the heavy borrowings of 1853, these sums averaged nearly $2,000,000 a year.

Expenditures in the later years included two notable cases of state works, each intended to overcome a mountain barrier considered too formidable even for generously-aided mixed enterprise. When the Louisa (later Virginia Central) Railroad Company found the Blue Ridge too much of an obstacle, the Board of Public Works tunneled the mountain and built some eighteen miles of track at a cost of about $1,000,000 and then turned its road over to the company for operation. The more important case was that of the Covington and Ohio Railroad which was to cross the second line of mountains and complete the connection between the James River and the Ohio which the James River and Kanawha had failed to make. More than $3,000,000 were spent on the construction of "this great national highway," but the work remained unfinished at the outbreak of war.[81]

In the administration of the extensive state program, both legislature and Board continued to follow the general policy laid down by the 1816 committee of reliance on individual initiative and limitation of state control. Practices developed for a system in which the state held a minority interest remained largely unchanged when the financial basis of mixed enterprise was altered. The state's vote in stockholders' meetings was deliberately held to a minority after its standard subscription became three fifths instead of two fifths, and after the state supplied more than half of the capital for the James River and Kanawha Company. The annual reports and still more the letter books of the Board of Public Works, with their instructions to the state proxies and state directors, show its self-restraint in the exercise of its power and its great reluctance "to obtrude upon the detailed administration" of the various companies. An extreme case was that of two state-aided railroads which quarreled over the interchange of freight until one railroad built a line parallel with the other. The Board deplored the dissension but made no attempt to use the state's power as a stockholder to compel agreement.[82] At the end of the period, the state was building the Covington and Ohio Railroad on the same route on which the James River and Kanawha Company—with legislative approval—was still planning an all-water connection.

There were, however, points at which the Board exerted firm pressure. From the beginning it insisted, with increasing and substantial success, on full annual reporting on the part of all improvement companies. The Board also provided important engineering services and where necessary insisted on their use. Its Principal Engineers, including Loammi Baldwin of Massachusetts and the Frenchman, Charles Crozet, brought new technical standards to replace what Crozet called the crudities of "the practical man." In 1831 the Board resolved to require a survey and location before payment of the subscription, and state proxies and directors were instructed to vote against departures from plans laid down by the Principal Engineer.[83] By these means and by checking occasional scandals, the Board exercised important functions of surveillance and oversight.

In a system frankly based on the principle of localism, it is not surprising that state money was spent on certain improvements that were really "limited to neighborhood convenience" and that the competition of communities for "a little of the sunshine of public favor" led to jealousies and charges of favoritism.[84] A legislative committee declared that Virginia harbored more local and discordant interests than any other state. A classicist from the Tidewater region charged the people of the West with wishing to make some splendid "Appian or Flaminian way" at the expense of eastern taxpayers,[85] and the men from the mountain counties in turn complained that most of the state's money was spent east of the Blue Ridge. Without authority to impose a general plan, the Board nevertheless exerted its influence to prevent the dissipation of energies on improvements that did not fulfill what it considered the original purpose of connecting remote parts of the state. For this purpose it included in the questionnaire to be filled out for each project the heading, "Links of Connection." In 1838 it declared that the state map prepared under its direction would serve to call attention to the "chasms" that remained to be filled in. In its 1851 report the Board claimed that a coordinated network of railways was in fact emerging, of which the principal elements were the line of roads from the Potomac to the North Carolina border and four routes extending from it to the west.[86] Of these, the Virginia and Tennessee, making the southwestern connection, was completed within the following decade.

In addition to the contributions of the state government, substantial sums were invested in internal improvements by Virginia local governments. The counties, which had had the principal responsibility for road building in colonial times, were active in encouraging the building of turnpikes, and a number of state-built or state-aided roads were later transferred to their jurisdiction. Two counties on the upper Ohio made railroad stock subscriptions of $650,000. Of the municipal subscriptions, those of Richmond and Lynchburg to the James River and Kanawha Canal, and of Wheeling to the Baltimore and Ohio Railroad, have already been mentioned. Alexandria subscribed $450,000 to the Alexandria Canal Company and invested another million dollars in railroads. Petersburg made an even larger commitment to railroads in its vigorous efforts "to add to the trade of the town." The municipality provided part of the capital and much of the impetus for the Petersburg Railroad, one of the earliest in the state, which extended southward to tap the Roanoke River trade and ultimately become part of the main north-south line. In 1847 the town bought out both the state and private interests in the City Point Railroad. In 1849 the state, recognizing Petersburg's "energetic and enterprising spirit," turned over its stock in the Petersburg Railroad "to the common hall of the town" for use in constructing still another railroad, the Southside.[87] According to the figures of Milton S. Heath, the total municipal aid to railroads in this period was $6,302,000 and the total of county aid $1,625,400.[88]

The records of the state Fund show a total holding in improvements in 1861 of more than $39,000,000, of which about $6,500,000 was in state works. If allowance is made for earlier investments that had been disposed of by the Fund and for the state's guarantee of the improvement bonds of companies and municipalities, the total of state aid and commitment in the construction of internal improvements was at least $43,000,000.[89] Taking account of the contributions of local authorities, the total of public money and credit committed to the improvement of transportation facilities cannot have been much less than $55,000,000, a substantial sum for a commonwealth of planters and farmers.

On this large outlay, the financial returns had been far from satisfactory. For the entire period from 1816 through 1861, the state's current expenditures on its investment program, including interest, exceeded revenue from improvements by almost $22,500,000, a deficiency made up by current appropriations and revenue from bank investments. Obsolescence had taken a heavy toll on the earlier improvements. Only one navigation company was paying dividends in 1860, and a committee of the House of Delegates calculated the rate of return on the state's investment in "minor improvements, including turnpikes, plank roads and bridges" as .007 percent.[90] On the other hand, half of the total investment and much more than half of the railroad investment were less than eight years old in 1860, and the state's annual income from railroads was over $300,000 and evidently on the increase. The Board of Public Works published annual tables of "railroad progress," and miles in operation had risen to 1,350. Instead of showing disillusion in the face of the figures of return, the legislature of 1860–61 authorized the largest appropriations in the history of the program. Thus Virginia ended the antebellum period, not in revulsion against state participation in internal improvements, but with actions that appeared to reflect pride in its accomplishments and confidence in the soundness of its system.

The successes of the Virginia system were largely in the more settled areas, in which at least some "individual wealth" could be "elicited for public improvement." Its greatest shortcoming was the failure to overcome the Appalachian Barrier. In the days before the railroad, the Kanawha Turnpike won only a small part of the trade of the Ohio Valley; [91] and war intervened before the completion of a railroad across the mountains to rival the Pennsylvania or the Baltimore and Ohio, which reached the river on Virginia soil but belonged to what Virginians regarded as "the Baltimore drainage." In spite of the prestige of the great names of Washington and Marshall, in spite of repeated warnings that the commonwealth was being "circumvallated" by more enterprising neighbors,[92] and in spite of the added argument in the sixties for an east-west route that would not pass over "a foot of abolitionist soil," [93] ante-bellum Virginia never succeeded in making effective use of the James-Kanawha gateway. If part of the explanation lies in the persistence with which the leaders of the company clung to their belief in an all-water route, another and significant part rests in the inherent contradiction between a system of orderly grants-in-aid to local enterprise and the concentration of funds and effort required for the accomplishment of a major developmental work.

The fourth of the possible mountain gateways suggested by Gallatin was to be found between the Tennessee and either the Savannah or the Santee. This pairing of rivers differs sharply from the other three both in its geography and in its transportation history. In each of the more northerly cases, the eastern river, the Susquehanna or the Potomac or the James, breaks its own way through the Blue Ridge; but in this fourth case the headwaters of the Savannah and of the entire Santee system lie to the east of the mountains. Though the tributaries of the western river, the French Broad and the Little Tennessee, cut their way through the western part of the mountain mass, the steep and unbroken eastern wall of the Blue Ridge presented a particularly formidable obstacle to internal improvement. No railroad crossed the Appalachian Barrier in this region until after 1880 [94] and it has never provided a major east-west artery of transportation. Nevertheless both North and South Carolina included plans for building across these mountains in their ante-bellum programs of internal improvement, and the latter attracted particularly widespread attention by its ambition to draw to Charleston the trade of the Middle West and thereby make it "the New York of the South." [95]

In colonial times, the South Carolina legislature assumed occasional responsibility for river improvements by the appointment of commissioners who were authorized to commandeer the services of local residents; in 1785 a more general plan of river improvement was outlined.[96] The Santee Canal, which connected the state's principal harbor with its principal river system, was built between 1793 and 1800 by private investors, who received from the state only the privilege of conducting a lottery to raise funds for the enterprise. The purpose, in the words of its chief engineer, was "to give the Inhabitants of the Upper Country an Opportunity to receive a Reward for their Industry, which would at the same time benefit the Metropolis." [97] Gallatin's Report cited this improvement along with the Middlesex as the country's two major achievements in canal building; but the Santee Canal, although kept in operation until the eighteen-fifties, was far from duplicating the financial success of the Massachusetts enterprise.

The beginnings of a state system of internal improvements date from 1817 with the establishment of the office of Civil and Military Engineer, the purchase of a private company which had attempted improvements on the Catawba and the Wateree, and the subscription to seven of the twenty shares of the Winyaw and Wando Canal Company. In the following year the legislature voted an appropriation of $1,000,000 to be spent at the rate of $250,000 per annum. The projects included the improvement of almost all of the state's many rivers, both above and below the fall line, together with sections of intercoastal waterway and a turnpike from Charleston to Columbia. Total state expenditures through 1834 approached $2,000,000, of which about $1,250,000 was for canals; and the work done included at least some improvement on nearly 2,000 miles of waterways and the construction of 25 miles of canals and 147 miles of roads. After 1819 the work was directed by a Board of Public Works and later by a Superintendent of Public Works. Their efforts, according to a contemporary, were "ably seconded by the personal exertions of gentlemen of fortune and influence," who took part in the superintendence and contributed the labor of their slaves in the slack season.[98] But the net operating returns on these enterprises never approached the interest charges, and the contributions to development as well as the financial results were on the whole disappointing. A legislative committee attributed part of the difficulty to the diffusion of effort over such a multiplicity of projects; and a disillusioned governor declared that there was hardly a public work in the state, except the State Road and the Columbia Canal, that "would find a purchaser . . . at public auction." [99]

By this time a more novel and a more successful undertaking had been completed by the method of mixed enterprise. The South Carolina Railroad and Canal Company, which was chartered in 1827, completed in 1833 what was then "the longest railroad in the world," running from Charleston to Hamburg on the Savannah River, a distance of 136 miles. Its locomotive, the "Best Friend of Charleston," was the first built in the United States for railroad use.[100] The name was a reflection of the road's primary purpose to divert the cotton trade of inland Georgia and South Carolina that would otherwise go down-river to Savannah. Total cost to the date of completion was about $950,000. The major backing was provided by merchants and other private individuals in Charleston; but this was supplemented by a stock subscription of $20,000 by the City Council and a loan of $100,000 from the state. When the returns on the road proved disappointing in the early years, this loan was extended for an additional ten years as "a liberal indulgence" toward a company "which risked so much in a problematical project." [101]

Meanwhile Charleston merchants had begun to look much further afield in their attempts to increase their city's trade. The new proposal was for a railroad over the mountains from Charleston to Cincinnati. "What a bold conception!" exclaimed the Charleston *Courier*. "What a magnificent project!" [102] Its effect would be to give Cincinnati its nearest port on the Atlantic and to make Charleston "the Commercial Emporium of the South." Southern rice and cotton would be exchanged for western meat and grain,[103] and Charleston would become the great outlet for the export of the produce of the Middle West. A political argument supported the commercial one; and Senator Robert Y. Hayne told his fellow Carolinians that the trade connection would give Middlewesterners a better understanding of institutions, including slavery, and an interest "in defending and maintaining them." [104] Meetings of interested citizens assembled in the proposed terminal cities and railroad conventions were held at points on the projected route—Asheville, North Carolina, on the French Broad, and Knoxville on the Tennessee. On the last of these occasions, Senator Hayne offered as a toast: "The South and West—We have published the banns." [105]

To carry out the project, South Carolina granted a charter in 1835, and supporting action was taken in the legislatures of North Carolina, Tennessee, and Kentucky. Since the last of these imposed the condition of a branch to Louisville, the company became the Louisville, Cincinnati, and Charleston Railroad. Funds were difficult to raise, and very few private subscriptions were made outside South Carolina. A further appeal to the legislature in 1836 brought substantial aid: a state subscription of $1,000,000 and the privilege of establishing a banking corporation whose stock could be held only by the stockholders of the railroad. The new provisions required the raising of a total capital of $8,000,000 before validation of the state subscription or organization of the South Western Rail Road Bank. The prospect of banking profits was an attractive one, but the canvass for subscribers coincided with the financial crisis of 1837; and the requisite figure was reached only with the aid of subscription of $650,000 by the state of Tennessee and $50,000 by the city of Charleston and by counting the assets of the existing South Carolina Railroad and Canal which the new company absorbed through an exchange of stock.[106]

After these beginnings the later history is anticlimactic. On the basis of surveys in the mountain region, the directors adopted the more easterly route by way of the French Broad River, though Calhoun and others preferred a line to the Little Tennessee.[107] Since no railroad even approached the mountains from the Carolina side, a sensible first step, once the South Carolina Company had been acquired, was to build a branch from its line to Columbia. Construction began in 1838 and the branch was completed in 1842 at a cost of about $2,250,000. Subscriptions were hard to collect in a time of deepening depression, and this effort exhausted the remainder of the company's funds. Long before the branch was finished, the great ambitions of a mountain crossing were abandoned as impractical. Tennessee and North Carolina subscriptions were returned, and in 1843 the new corporation gave up its more grandiose title and became the South Carolina Railroad. What it had accomplished was the completion of a useful and ultimately profitable rail connection between Charleston and Columbia, at a time when construction in many states had come to a complete standstill. What it had abandoned was the attempt to overcome the mountain barrier.

The challenge was renewed in the eighteen-fifties. By this time railroads in the Piedmont gave a base from which to enter the mountain country, and the proposal was to make the crossing by a more westerly route between the headwaters of the Savannah and the Little Tennessee near the point at which the three states of

North and South Carolina and Georgia meet. For this purpose the Blue Ridge Railroad Company was chartered in 1852 with subsidiaries in the other three states concerned. Private subscriptions were very much smaller than for the earlier project; and almost the entire financial support came from the state of South Carolina, which subscribed $1,000,000 and endorsed $1,000,000 of the company's bonds, and from the City Council of Charleston, which took $1,049,000 of the company's stock. The Blue Ridge Railroad confronted from the outset the difficulties of building through mountain terrain, and plans called for the digging of a tunnel of nearly 6,000 feet through the solid granite of Stump House Mountain. Work progressed very slowly, in part because of troubles with unsatisfactory contractors; and an additional subscription of $300,000, made by the state in 1858, was soon spent. When work was finally abandoned in 1861, thirty-three miles of track had been laid, and the tunnel had been carried with great effort two thirds of the way through the mountain.[108]

By contrast a network of less ambitious railroads was completed between 1847 and 1861 by a combination of private and public effort. A number of short roads in the Piedmont were started by local subscriptions, with some town aid. The city of Charleston subscribed $400,000 for a road to connect with Wilmington, North Carolina, and $260,000 for a road to Savannah. State aid was added in almost every case but without attempt to influence the planning or choice of routes. Subscriptions to a dozen companies were made from a "securities revolving fund" based originally on the state's holdings in the South Carolina Railroad Company and the South Western Railroad Bank. In this way a total of state subscriptions to railroad companies amounting nominally to nearly $4,500,000 was made at an original cost to the state of a little more than $2,500,000.[109] The state also guaranteed railroad bonds totaling $4,680,000. Together with the small amount of loans, this brings the figure of state participation and commitment to well over $9,500,000. In the same period municipal aid amounted to a little more than $4,000,000, of which Charleston provided all but a small amount.[110]

South Carolina built one of the country's first canals and one of its first important railroads. Yet its public effort in internal improvements, in spite of the substantial contributions of the city of Charleston, was on a lesser scale than in the states previously discussed. During the railroad period, there was no state organ attempting the role of central study and surveillance performed by the Virginia Board of Public Works. On the other hand, private railroad investment, beginning with the Charleston and Hamburg, was of substantial amounts. In 1861 the total capitalization of the railroads was $22,000,000, and about 700 miles of road were in operation. There were no defaults before the Civil War on the bonds endorsed by the state, and the workability of the state's revolving fund was an indication that the earlier lines had become substantial properties. As for the "transmontane" railroad that was to surpass the achievements of Xerxes and Hannibal,[111] the unfinished tunnel in Stump House Mountain was evidence of complete failure. Ante-bellum South Carolina had been unable either to amass the resources needed for so great an undertaking or to arrive at a clear recognition that the project was economically impracticable.[112]

The western counties of North Carolina lay across both of the proposed routes between the Tennessee Valley and the eastern waters. Though the state's ports were inferior to those of Charleston and Savannah and further from the mountains, there were nevertheless patriotic North Carolinians who argued that their territory provided the best of all natural routes from the Atlantic to the West. Proposals for a great central railroad extending from the seacoast across the mountains competed for attention with more local projects and with lines crossing the state from north to south.[113] Public support for these various undertakings was given on a scale comparable with that in South Carolina;

but private investment was much less, since there were no accumulations of wealth to match those of the Charleston merchants, and the major part of the burden was assumed by the state government.

Early efforts in North Carolina resembled those of Virginia and South Carolina in philosophy and organization. Between 1815 and 1819, a series of committee reports and pamphlets prepared by Archibald D. Murphey outlined a comprehensive plan of measures to improve navigation on the state's rivers, to build highways in the mountain country, and to improve the inlets in the sandy barrier that separates the ports and sounds of North Carolina from the Atlantic. The first of these proposals included the principle of state subscription of one third of the stock of navigation companies formed for each of the state's primary rivers. The entire program included appointment of a Board of Internal Improvements and, "as a primary part of the Plan . . . the employment of a Civil Engineer of Science and Experience to direct this great work." [114]

The plan was not adopted as a whole, but state subscriptions to the Roanoke and Cape Fear Navigation Companies were made in 1815. Four years later, the services of an experienced English engineer were secured and the Board for Internal Improvements was organized.

"The leading objects of a system of Internal Improvements in North Carolina," said the Board in its 1821 report, "ought to be

1. To give to all our citizens an easy and commodious way of getting to market the productions of their industry.
2. To fix that market within our own limits." [115]

The second of these purposes expressed a concern, like that of the Philadelphians in Pennsylvania, with the fact that so much of the state's produce found its markets in the neighboring states. Most of the trade of the Roanoke Valley went to Norfolk either overland or by way of the Dismal Swamp Canal, and much of the trade of the central part of the state followed the river valleys down into South Carolina. It was therefore argued that construction of a canal from the Catawba and the Yadkin to the Cape Fear to divert these shipments would have "great national"—i.e., state—importance. If these leakages to north and south could be checked, said the Committee on Inland Navigation, North Carolinians "might indulge the hope that a commercial city would grow up" on their coast "equal in importance to Philadelphia, Baltimore and Charleston." [116]

The results fell far short of these aims. The state subscribed to the stock of companies organized to improve the navigation of six rivers, to attempt a canal between the Yadkin and the Cape Fear, and to cut a short intercoastal passage. A number of these subscriptions were made on condition that the companies "place the execution of their works under the direction of the Chief Engineer for the State, and the superintendence of the Board for Internal Improvements." Except on the Roanoke, for which Virginia's subscription was larger than North Carolina's and to less extent on the Cape Fear, little progress was made; and most of the companies abandoned their efforts. The failure of the program has been attributed partly to blunders made "before the aid of science and skill had been enlisted to direct the operations" and partly to the diffusion of effort among so many projects.[117] In any case the entire effort was on a very small scale, with meager private investment and a total state expenditure to 1833 of less than $300,000.

North Carolina's railroad effort was much more substantial. A proposal for a great east-west railroad was made as early as 1828, in the Reverend Joseph Caldwell's "Numbers of Carlton"; and an internal improvement convention, held in November 1833, framed a comprehensive program of improvements, mainly railroads, designed to serve the interests of the various parts of the state. It was proposed that these be built as state works at an estimated cost of $5,000,000. The first actual railroad construction, however, which began in 1836, was done on two north-south lines initiated by private enterprise.

One of these was the Wilmington and Raleigh, later the Wilmington and Weldon, built—as its second title indicates—to connect the state's chief port with the Virginia railroads on the northern border. Though the work was begun without public aid, subscriptions were hard to raise; and in 1837, the state, enriched by the windfall of the federal surplus revenue, agreed to subscribe two fifths of a capitalization increased to $1,500,000 and took two of the five places on the board of directors. During the difficulties of the depression years which followed, the state made possible the continuation of the work by endorsing $300,000 of the the company's bonds. The line of more than 160 miles was completed in 1840. By 1845, net revenues had begun to exceed interest charges and dividends were paid regularly from 1851 on. In the latter year the state transferred $200,000 of its stock in the company to aid a connecting road, the Wilmington and Manchester, which continued the line to the southward with additional assistance from the city of Wilmington and from the state of South Carolina. The Wilmington, Charlotte, and Rutherfordton also received state encouragement, in the form of a bond endorsement of $200,000 for each twenty-five miles; and built northwestward into the Piedmont for more than a hundred miles, crossing the lines of trade to South Carolina.

The second of the original pair of railroads, the Raleigh and Gaston, was also begun in 1836 and completed in 1840. A shorter road with a more checkered history, it ran the eighty-five miles from the state's capital to the Roanoke River, not far from Weldon, to connect with the Petersburg Railroad. Private subscriptions along the route and in Virginia provided inadequate capital and the road received no windfall from the surplus revenue. The legislature was induced to endorse $500,000 of the company's bonds in 1839 and an additional $300,000 in 1841.

On the latter occasion it took as security not only a new mortgage but a personal bond of $500,000 given by the stockholders and other individuals. Earnings failed to cover interest or even proper maintenance. The company defaulted on its interest payments in 1843, the personal bond proved uncollectible, and the state foreclosed in 1845. For six years the road was run as a state enterprise. Under a reorganization in 1851, the state retained a half interest valued at $400,000 and invested an additional $87,500 to cover half the cost of a branch line from Gaston to Weldon. By the mid-fifties the Raleigh and Gaston was a dividend-paying road; but the amount still due on the principal of the old endorsed bonds, plus payments already made for principal and interest and a small operating deficit incurred during the period of public management, exceeded the state's equity by more than $900,000.

Neither of these companies met the demand for a great central railroad, and the Raleigh and Gaston ran directly counter to the ideas of the home market advocates by providing more efficient means for carrying North Carolina products to Petersburg and Norfolk. A second wave of railroad investment, in which the state provided the major part of the funds, was devoted to a series of lines following in general an east-to-west direction. The first of these was the North Carolina Railroad, chartered in 1849 to run from Goldsboro on the Wilmington and Weldon through Raleigh to Salisbury and Charlotte. The state agreed to subscribe two thirds of the capital stock of $3,000,000 but only after $1,000,000 of private subscriptions were made and half of the sum paid. It took the promoters two years to meet these conditions; and the selection of a somewhat roundabout route from Raleigh to Salisbury, which was made before the state was represented on the board of the company, suggests the importance attached to collecting subscriptions in each of the towns along the way. Even so, the $500,000 required to validate the state's contribution in 1852 was obtained only by inducing the contractors to begin the work and take payment in stock. The state subscribed a further $1,000,-

000 in 1855, bringing its ownership to three fourths. With this assistance and the flotation of a small bond issue, the company completed the road in the following year. Traffic was substantial from the start; and on the eve of the Civil War, though no dividends had yet been paid, operating revenue had risen to almost five percent on the original investment.

The dream of a great central railroad still required extensions to the seacoast and across the mountains. To accomplish these purposes the legislature agreed in 1855 to subscribe for two thirds of the stock of the Atlantic and North Carolina and the Western North Carolina railroads.

The former was to build from Goldsboro via New Bern to the harbor of Beaufort. Since the capitalization was set at $1,600,000, this required the promoters to raise $533,400. They found few private subscribers, and the initial list fulfilling the requirement included $250,000 in county subscriptions and nearly $200,000 set to the account of contractors who agreed to accept payment partly in stock. By the time the bulk of the money had actually to be secured, the problem was complicated by the controversy over the location of the ocean terminus. The decision to place it on the western side of Beaufort Harbor rather than on the eastern side, on which the town of Beaufort was located, had the effect of sacrificing a subscription offered by the town, and jeopardizing that of Carteret County, but of obtaining substantial support from the proprietor of the new real estate development of Morehead City. Additional state aid was obtained in 1857 in the form of a $400,000 loan.[118] Actual construction offered no great difficulties, and the road was finished in 1858. The period of operation before the Civil War was too short to provide a test of the road's earning power, but there was little to suggest that it would turn Morehead City into a major port.

The Western North Carolina undertook an enterprise that was geographically and financially much more formidable. Its purpose, as stated in the act, was to form a connection with the

Mississippi Valley; and the legislature committed itself to providing $4,000,000 of a total capitalization of $6,000,000, with the important condition that the road was to be done in sections and the work completed on each section before any state money could be applied to the next. The line was to be carried westward from Salisbury on the North Carolina Railroad across the Blue Ridge to Asheville on the French Broad and beyond as the legislature might later direct. The first section was to run to Morganton at the eastern base of the Blue Ridge. Private subscriptions sufficient to validate a state subscription of $800,000 were secured and the company was organized in 1855. Three years later, with the aid of a Burke County subscription of $50,000 obtained by a close margin after an earlier proposal had been defeated by the voters, another $220,000 was raised and matched by $440,000 more from the state. Outbreak of war found the first section still five miles short of its terminus at Morganton, but even the uncompleted line was providing a considerable amount of traffic for the North Carolina Railroad.

In the meanwhile Western North Carolina had been agitating for the immediate commencement of the work across and beyond the mountains. Some of its spokesmen wished the line to run into Tennessee down the valley of the French Broad, and others preferred a route through the extreme southwestern corner of the state in the direction of Memphis, but both factions demanded the elimination of the section rule so that state money could be made available to match subscriptions for the western portions of the road. The Asheville *News* hailed the example set by the Burke County subscription: "We trust McDowell will next put her shoulder to the wheel, and shove it along to the Buncombe line, when, our word for it, the noble old 'mother of counties' will not prove a laggard in the work." [119] McDowell did in fact vote a $50,000 subscription; but when the 1859 legislature, after a bitter contest between eastern and western interests, refused to eliminate the section rule, Buncombe and two neighboring counties voted to subscribe instead to a project for a South Carolina connection. In spite of these discouragements, a considerable amount of grading and other work was done on the summit part of the route, and late in 1861 the state finally released $220,000 of its subscription for payment of the contractors.

At the outbreak of the war, North Carolina was closer to a successful crossing of the Blue Ridge than South Carolina, and a state with very limited resources had built some 850 miles of railroad. For their construction, the state government had spent or committed more than $10,000,000; and smaller amounts had come from private, county and municipal subscriptions.[120] The earlier program of state aid to roads and inland navigation was under the superintendence of a Board for Internal Improvements, but the much larger expenditures for railroads were made without central direction. The advocates of a concerted plan placed their emphasis on a great east-west route from the coast to a crossing of the Appalachian Barrier. The state's actual investment, however, showed no consistent pattern, and the results reflected the strength of the economic forces stimulating trade along lines paralleling the seaboard and extending southward from Washington to South Carolina and beyond. The most successful of the state-aided railroads was the Wilmington and Weldon, which crossed the state from north to south and was later joined with its Virginia connections to form the Atlantic Coast Line Railroad. Support of the Raleigh and Gaston, which took so much of the public funds, appeared "suicidal" to those who thought that the prime object of state policy should be to bring trade to North Carolina ports.[121] Well over half of the state's investment, to be sure, was in the three roads which formed a connected route from the coast to the interior; but of these the North Carolina Railroad was built on such a "sweeping curve" that much of the line was later to find its greatest usefulness as part of the north-south route of the Southern Railway.[122] There was indeed little reason to believe that a "great central railroad" could ever have fulfilled the dream of drawing streams of produce from the interior to a new Philadelphia behind the sandy inlets of the North Carolina coast.

In Georgia the mountains were lower, as the Appalachian Barrier approached its end, and they presented a less formidable barrier to transportation. It was therefore possible, in Calhoun's phrase, to "turn the Alleghany to the Southwest, as New York had to the Northwest." [123] Early projects called for a canal to connect the Tennessee with the river system of Georgia, which indeed Gallatin had suggested; [124] but the agency which finally turned the barrier and formed the connection with the Middle West was the Western and Atlantic Railroad. Like New York's Erie Canal, the road was an enterprise of the state government; and it bore a unique and planned relationship to the other internal improvements of the state, which were mainly the work of private individuals and city governments.[125]

In 1804 and 1805, Georgia built a state wagon road to the Tennessee line through what was then the Cherokee country, and there were various early efforts at river improvement, sometimes carried on as in Virginia by voluntary organizations and occasionally receiving small amounts of state aid. After appropriations were made for two rivers in 1815, a more comprehensive program of river improvement was initiated in 1817; and a permanent Internal Improvement Fund of $250,000, later increased to $500,000, was set up in the same year. State encouragement was also offered to several turnpike companies, in the form of stock subscription, but none were successfully organized. Demands for a more concerted approach to the problem led to the establishment of a Board of Public Works in 1826, much later than in Virginia and the Carolinas. As a member of this Board, Wilson Lumpkin examined routes for the proposed canal to the Tennessee River and selected the same gap between Lookout Mountain and the main range of the Blue Ridge that had been used by the early road and was eventually to be used by the Western and Atlantic. The Report of the Board and of the State Engineer, presented the following year, laid greater stress on the importance of railroads and presented a notable analysis of Georgia's transportation needs and possibilities. It argued for "a state system of transportation" that would on the one hand resist seduction "by the siren magic of distant trade" which had been so persuasive in South Carolina and on the other hand avoid the wastefulness of "uncoordinated local efforts." [126]

The Board was abolished after a year's service, largely because of sectional rivalries, and the state returned to more piecemeal efforts at river improvement and road building. Much of the work was done by slaves, the so-called "public hands," that had been purchased by the state. A total of over $400,000 was spent on river improvements to 1840, with disappointingly little result, and a smaller amount on the construction of market roads. Agitation for the building of a central canal had been checked by the Board's comments on the advantages of railroads, but two smaller canal projects were actively pursued. Each was planned to connect a port with the state's largest river system—one to run from Savannah to the Ogochee and thence to the Altamaha, and the other to run the shorter distance from the harbor of Brunswick to the Altamaha. Though the former received aid from the city of Savannah and nearly $100,000 from the state and the latter received $50,000 from the state, Savannah's canal reached only its first river and never the second, and the Brunswick canal was dug but never opened to traffic.

Railroad building began with two campaigns of private and local effort. The first, originating in the towns in the interior of the state, resulted in the construction of the Georgia Railroad westward from Augusta on the Savannah River. The second resulted in the building of the Central of Georgia Railroad from Savannah to Macon in the center of the state and of several connecting roads radiating from Macon. The two principal companies were chartered in 1833 and began construction shortly thereafter. The Georgia road received a small county subscription and also $200,000 from the city of Augusta on the embarrassing condition that the railroad should not connect its line with the Charleston and Hamburg without permission of the city council. The Central of Georgia received more substantial municipal aid. Savannah subscribed $500,000, together with $50,000 to a connecting road, and Macon subscribed $250,000. Neither received a financial contribution from the state, but the grant of banking privileges by an act of 1836 stimulated private subscription, particularly in the case of the Central. With work continued during the depression years, the Georgia Railroad was carried as far as Madison and Athens by 1841, and the Central of Georgia was completed to Macon in 1843.

Meanwhile the prospects of these companies, and of Georgia railroads in general, had been influenced by a decisive new development in state policy. Acting on the recommendation of Governor Lumpkin, the legislature of 1834 adopted a resolution in favor of "a scheme of internal improvements from the seaboard of this state to the interior by railroad on the faith and credit of the State as a great State work." [127] What developed was a state enterprise more limited in scope but more unusual in character. Under the powerful advocacy of Governor Lumkin and the officials of the existing railroad companies,[128] the project was elaborated in the Macon Railroad Convention of 1836 and adopted by the legislature in the same year. It provided that the state should on its own account build a railroad through the mountainous northwestern part of the state from a point south of the Chattahoochee River to the Tennessee border near what is now Chattanooga. From the southern end of the road, a locality referred to as Terminus before it received the name of Atlanta, branch lines should be built to connect with the remaining parts of the state. These branches were not themselves to be state enterprises, though on certain conditions the state might subscribe to a quarter of their stock; they were to be built either by the existing railroads or by other companies organized by private and local enterprise.

The effect of this decision was to assign to the state government the more clearly developmental part of the task of railway construction: it was to build in the mountainous country where the costs would be the greatest and the prospect of immediate traffic the least. If the vital connection with Tennessee could be achieved by this means, companies organized on the basis of mixed enterprise could do the rest. The result would be a coherent network of transportation extending to the coast and throughout most of the state. With "the main trunk . . . common in its use," declared the Macon Convention, "this system of railroads would diffuse over all parts of the state almost a precise equality of advantages." [129]

Construction of the Western and Atlantic proceeded slowly under depression conditions, even though contractors accepted payment in state scrip, and an economy-minded administration suspended work from 1841 to 1843. The line was not completed to Chattanooga until 1850. The extension of the Georgia Railroad reached Atlanta in 1845, and the Central of Georgia's connection, by way of the Monroe Railroad, was established in the following year. The southwestern prongs of the system, to Columbus and other points on the Alabama border, were built later and with greater difficulty. Early returns on the state road were small, and the legislature was at one time prepared to sell the property at a substantial loss; but revenues rose steadily during the fifties, and the annual payment to the treasury after debt service reached the figure of $450,000 in 1860. Since total appropriations to that date were about $4,500,000 and an additional $1,600,000 had been reinvested from earnings, this return to the state—for the part of the railroad system assigned to it as the least attractive to private enterprise—represented ten percent on the sum appropriated and more than seven percent on the total cumulated investment.

In the latter fifties the policy of state investment, though not of state construction and operation, was revived for another trunk-line project for which somewhat similar arguments could be made. The south and southwest of the state were little developed, and a route through them would provide access to the Gulf ports of western Florida and Alabama. Rival companies to build in this direction were organized by Savannah and Brunswick interests. In 1856 the Albany and Gulf Railroad was chartered to construct a trunk line westward from a point of intersection of these two roads. The state subscribed $1,000,000 of a capitalization of $2,200,000, and most of the remainder was taken by the city governments of Thomasville and Savannah. By the outbreak of war, construction had been completed for the greater part of the route across the state. Determined efforts to secure state aid for a number of other railroads were defeated in the legislature, and the state's financial contribution to railroad development was confined almost exclusively to the two trunk-line projects.

Total state investment in the railroad program was about $5,500,000 in initial aid, and more than $7,000,000 if the reinvestment of earnings is included. As Heath has shown, the sums devoted to railroads, aside from debt service, amounted to about a quarter of the state's total expenditures in the eighteen-thirties and over a third in the forties. For 1851–60, they were still eighteen percent of the budget but a little more than matched by the revenue from the Western and Atlantic.[130] Local government investment, mainly by stock subscription, amounted to some $5,750,000. In this the municipalities took a much larger role than the counties, and Savannah's share was nearly $2,750,000. The city's investments were first placed in the Central of Georgia and in aiding the same road to encourage the building of a network of lines from its Macon terminus. Later in the period they were applied, partly by use of the dividend-paying Central stock as a revolving fund, to the encouragement of railway building toward the southwest and the Gulf. Though the city authorities seem to have taken little part in the direction of the companies which they aided,[131] their strategic role in the placement of investments was almost comparable to that of the state itself. On the other hand, the total of private investment in Georgia's railroads, plus the reinvestment of the earnings of mixed enterprise, fell just short of equaling the public share.

By these means ante-bellum Georgia, which had had little success with its earlier waterway improvements, secured a network of some 1,200 miles of railroad. Georgia's mileage was roughly equal to that of Virginia, which had committed well over twice as great an amount of public funds, and considerably greater than that of South Carolina, which had spent somewhat more public money and made a much larger private investment. As compared with the Carolinas, Georgia had the very great advantage of a practicable route through the mountain barrier. Yet the decisiveness with which the geographic opportunity was seized, and the coherence of the railroad system which was developed, owed much to a public policy which so largely concentrated the effort of the state government on a single strategic improvement.

4

FROM BOSTON TO MOBILE

The Governor of Missouri boasted in 1850 that railroad promoters all the way "from Boston round to Mobile" were competing for the privilege of building to his state and through it to the Pacific.[1] Even this phrase is not broad enough to cover the full range of public action for the development of transportation during the period before the Civil War. The line of coastal cities could be extended to reach from Bangor to Galveston and beyond; and in almost every state from Maine to Minnesota and from Florida to Texas governmental agencies—state or local and more often both—took an active part in the promotion of internal improvements. For the non-Appalachian states there was no such common geographic goal as that which gave unity to the preceding chapter; but their programs show the same emulation of states and cities, similar calculations and miscalculations of advantage, and perhaps an even greater diversity of expedients in the combination of public and private effort.

Among the Atlantic states that had no direct access to the main Appalachian Barrier, there were several in which there was little reason to expect major public effort toward internal improvement. In Delaware there was perhaps more than its size and geographical position would suggest. The canal between the Chesapeake and Delaware bays, a project discussed even when the colony was in the hands of the Dutch,[2] was finally completed in 1829. Delaware's senators led the campaign for federal aid to the enterprise; but the state's own subscription of $25,000 was less than those obtained from the federal government and the states of Pennsylvania and Maryland. Four years after the opening of the canal, Delaware loaned its stock to the New Castle and Frenchtown Turnpike and Railroad Company. Another improvement clearly indicated by the state's geography was a railroad across the northern part of the state between Philadelphia and Baltimore. Though the connection between two such major cities might have seemed to offer an opportunity that could be taken by unaided private enterprise, the state government assisted the Philadelphia, Wilmington and Baltimore Railroad (which later became a part of the Pennsylvania system) with a loan of about $80,000. This, together with aid to several smaller roads, was provided for from Delaware's share of the surplus revenue of 1837. The largest amounts of state aid, reaching a total of $130,000 in stock subscription and a loan of $170,000, were given to the Delaware Railroad built through less wealthy areas "to the southern line of this state in a direction toward Cape Charles." Impatience with this mounting total of costs was shown by the legislature in the acts of 1855 and 1859, which made concessions to the company but attempted to provide that it should never again "ask for any pecuniary aid from the State or for the loan of State credit." [3]

The case of New Jersey approaches the minimum of direct public aid to transportation. The only money actually appropriated by the state for this purpose was an early subscription of $12,500, authorized in 1804, to half of the stock of the Newark Turnpike Company. The project languished for a number of years but eventually proved profitable. From 1818 on, the state appointed two of the five members of the board of directors. Indirect aid to two interstate turnpike companies was authorized in the form of lottery privileges, but the lotteries were conducted inefficiently or fraudulently, and little investment was actually made.

New Jersey's abstention from financial aid to improvements, after the turnpike period, has sometimes been attributed to the exceptional self-reliance of New Jersey's enterprisers or to the devotion of its legislators to the principles of laissez faire. Jerome Cranmer has recently made an attempt to determine "the peculiar forces or conditions which enabled (or forced) the State to provide for her two major internal improvements, the Morris and the Delaware and Raritan canals, without resort to State financial aid."[4]

In the second of these cases, the explanation runs in quite obvious economic terms. The route of the Delaware and Raritan, which was completed in 1834, lay across one of Gallatin's four necks of land; New Jersey parlance changes the metaphor and calls it the "waist" of the state. If this was the only one of the four to be exploited during this period by unaided private enterprise, it was the one that presented the special opportunity to connect New York and Philadelphia by a cut of some thirty miles through almost completely level country. The private company chartered in 1824, instead of receiving a subsidy, paid the state a bonus of $100,000 for the privilege of building the canal.

Yet the history of actual construction followed a more complex pattern. The original company failed, partly because it could not secure the consent of Pennsylvania to the necessary diversion of water from the Delaware River; and measures providing for the construction of the canal as a state work came close to passage in the legislatures of 1827 and 1828. By the time building actually commenced, the financial prospects had become less promising because of the innovation of the railroad; and the terms under which the canal was constructed formed the less important part of the set of arrangements under which the same intercity route was exploited by rail transport. Early in 1830 the legislature passed two measures on the same day, separate in form

the state was to subscribe a third of the stock, were decisively defeated in 1830. The Hudson project was renewed with the completion of the Boston and Worcester, which provided the first forty-four miles of its route. Private subscriptions amounting to $2,000,000 were obtained for the Western Railroad, in part by a house-to-house canvass of Boston; and the Legislature in 1836 reversed its earlier position and authorized a state subscription of $1,000,000. The crisis of 1837 and the high cost of construction through the Berkshires kept the company in financial difficulties, and the state came to its rescue with a series of loans amounting by 1841 to $4,000,000. With this aid work was continued during the depression years; and the entire two hundred miles of what is now the Boston and Albany—the Boston and Worcester, the Western, and a New York connection largely financed by the city of Albany—were open for traffic in 1842.

The Western Railroad proved profitable and repaid the state's advances. To Boston it brought an important addition to its western trade by tapping the routes of the Erie Canal and the New York Central Railroad, though it did not—and, considering grades and distances, could not—replace New York City as their principal beneficiary.

In the meantime, the depression conditions which had led to the loans to the Western resulted also in the granting of aid to a number of smaller railroads. Loans amounting to nearly $1,200,000 (all eventually repaid) were made to the Eastern and the Boston and Maine, which were building toward Maine; the Nashua and Lowell, which was building into New Hampshire; and the Norwich and Worcester, which was building into Connecticut. These appeals for help and a decline in the market for state bonds led, as in other states, to something of a revulsion against internal improvements; and the constitutional convention of 1853 came close to adopting a clause prohibiting state loans to private companies. This provision was defeated largely by the proponents of another intersectional improvement.

The new proposal was for a railroad to the West following the route which Loammi Baldwin had chosen for the projected canal. A pair of railroads, connecting at Fitchburg, crossed the northern part of the state to Greenfield on the Connecticut. Investors in Troy, in their eager rivalry with Albany, were ready to build to the Massachusetts line. What remained was to "construct a tunnel and railroad under and through the Hoosac Mountain," over which motorists today pass on the Mohawk Trail. For this purpose, the Troy and Greenfield was chartered in 1848. So formidable a project, which would serve the less settled and less wealthy part of the state, could be carried through only with the aid of public funds. Accordingly the Troy and Greenfield asked for a state loan. Its advocates reminded the legislature that "the votes of the northern border of the State . . . were united with the southern in granting the aid of the State to the Western Railroad. . . . Now," they said, "the time of reciprocity [had] arrived." [13] Against the proposal stood the interests of the Western Railroad and of Springfield and other towns along its route, as well as those who opposed state aid on principle or who thought that costs would outweigh benefits in building a second railroad parallel to the first. In its favor were the determined representatives of the northern section, "a part of the country for which nothing had been done," [14] and others who saw visions of a road superior to its rival and drawing an almost unlimited trade from the West.

In the 1854 legislature, as in the convention, the forces of the tunnel proved stronger than those of the Western: the state granted a loan of $2,000,000 to be made available in installments as certain conditions of subscription and construction were met. Even with this backing, little private money was ventured; and, when the state authorized subscriptions by the towns along the route, only half of them voted in the affirmative. By 1859, the paid-in capital was still considerably less than $400,000, and the company had been taken over by a contractor whose performance was a subject of violent controversy. In the following year the state liberalized the terms under which its funds were to be released, but work was stopped in 1861. More than a million of the state's money had been spent, together with $125,000 of the towns',[15] and about half a mile of the tunnel had been built with four miles still to go.

By the time of the Civil War, the state of Massachusetts had devoted some $7,350,000 to railroad promotion, and its towns had contributed $228,000. The total, though larger than that of the rest of New England, was considerably less than in any of the seven Appalachian states. The state's aid, as given in the eighteen-thirties and early forties, made possible the building of a useful intersectional railroad and served to keep alive other viable roads. But its latest venture, the Troy and Greenfield, stood in the position of South Carolina's Blue Ridge Railroad, with the difference that Massachusetts was left with a shorter tunnel dug into the side of a longer mountain.

Rhode Island and Connecticut had less occasion for ambitious projects. When the New York, Providence, and Boston was chartered in 1832 to build from Providence to the Connecticut line, the state reserved the right to subscribe to 1,000 shares, but no such purchase was ever made. The city of Providence, however, subscribed $500,000 in the late eighteen-forties to the Hartford, Providence, and Fishkill, which was projected to connect with the Erie and which went into bankruptcy a decade later.

In Connecticut, also, the state provided no funds for railroads, but there were several cases of municipal aid. Most characteristic were the efforts of its competing coastal cities to promote north-south railroads extending into Massachusetts. New Haven, disillusioned by its canal experience, made no contribution either to the New Haven and Hartford, which became the state's dowager railroad, or to the rival road which made use of the bed of the abandoned canal. On the other hand, Bridgeport competed eagerly in the late thirties to obtain the southern terminus of the Housatonic Railroad, subscribed $150,000 to its stock, and soon defaulted on its own bonds as a result. In the early forties, Norwich loaned $200,000 to the Norwich and Worcester; and in 1852 New London made a loan of $100,000 for its northern connection which struggled as far as Amherst before the Civil War and ultimately became part of the Central Vermont.[16] Meanwhile, Hartford had matched Providence's contribution to the east-west Erie connection.

Neither state nor municipal aid was given to railroads during this period in New Hampshire and Vermont,[17] yet a boldly conceived intersectional route was built across their lightly settled and rugged terrain. Its objective was to bring to Boston the trade of Canada and of the Great Lakes by a route far enough to the north to circumvent the attractions of New York City. Montreal was to be reached by way of the head of Lake Champlain; and Lake Ontario was to be reached at Ogdensburg, New York, in the hope that Canada's Welland Canal would draw to it much of the traffic of the upper lakes. The Vermont Central was to accomplish the crossing of the Green Mountains; its ally, the Vermont and Canada, was to provide the upper portion of the New England route; and the Northern of New York was chartered to complete the connection with Lake Ontario. From Boston the connecting route could begin with three profitable "lower roads" extending to Concord, New Hampshire, or could be made by other railroads in western Massachusetts and Vermont. In 1851 the last link was completed, with a floating bridge across Lake Champlain at Rouses Point; and trains could run from Boston to Lake Erie and the St. Lawrence.

The physical achievement was a notable one, but the economic and financial results were less impressive. The new railroads contributed to modest local development, and Boston did obtain by this route a share of the increasing trade between Canada and the United States; but the great hopes placed on traffic from the Lakes proved illusory. The financial history is a tangled one.

To complete the roads, the promoters pieced out insufficient subscriptions by resort to short-term borrowing or by farming out the enterprises to contractors paid mostly in stock. After completion, the record of the upper roads is one of bankruptcy, reorganization and—in the case of the Vermont Central—the country's "first major railroad receivership." [18]

In most states ambitious projects under similar circumstances received public aid. The restraint of New Hampshire and Vermont was in this period as complete and, considering their geographic situation, much more remarkable than that of New Jersey. It can hardly be attributed entirely to the reaction against public programs exhibited in a number of states during the depression years, since the upper roads were not launched until prosperity had returned. Part of the explanation, no doubt, lies in the rock-ribbed conservatism of citizens and legislators, as illustrated by the refusal of New Hampshire between 1840 and 1844 to allow to railroads even the right of eminent domain. But another part of the explanation must be found in the extraordinary faith which Boston investors, who provided the greater part of the capital, placed in so venturesome an undertaking.

An even bolder enterprise, and a much more successful one, had its origin in Maine. Here again the state gave no financial aid. Though Maine had established a Board of Internal Improvements in 1834, no action was taken on its recommendations, and in 1848 the state bound itself by constitutional amendment not to loan its credit to private enterprises. What the state refused to do was undertaken by the ambitious city of Portland.

The project was a railroad to Canada. The physical obstacles appeared even greater than those encountered by the Vermont roads. The White Mountains rather than the Green stood across the route, and much of it led through almost completely unsettled wilderness. Yet there were real economic advantages to which its advocates could point. "Portland," they said, "is the natural seaport of the Canadas." [19] The St. Lawrence was closed for the winter months, and Portland's ice-free harbor was nearer to Montreal than Boston and much nearer than the ports of the Maritime Provinces. Moreover, by the ocean speeds of the day, Portland was two days closer to Europe than New York and a half day closer than Boston.[20]

The Atlantic and St. Lawrence Railroad was chartered in 1845. Its principal propagandist, John A. Poor, traveled by sleigh through the storms of the same winter to persuade the capitalists of Montreal to place their money on a Portland rather than a Boston connection. As a result they organized the St. Lawrence and Atlantic to build to the American border. When private funds for the American company ran out, the city of Portland loaned it $1,000,000 in 1848 and by later enactments doubled the amount. This "First International Railway," which received public aid on both sides of the border,[21] was opened from Portland to Montreal in 1853. It was then leased to the Grand Trunk Railroad of Canada at a rental which made the city's investment secure. Before the end of the decade this provided Portland with a through route to Detroit, at a time when Boston's Western Railroad still reached only the east bank of the Hudson. By this alliance between its own private and municipal enterprise and Canadian capital, public and private, the city of Portland secured a notable increase in trade. Poor's newspaper, the *State of Maine*, had good reason to celebrate in verse the completion of

> That iron arm that links Atlantic "Maine"
> With Huron's waters, in a single chain.[22]

After this first example, the policy of municipal aid was continued in the development of Maine's other railroads. In 1850 the towns along the route of the Kennebec and Portland were authorized to loan their credit to the enterprise, to the amount of $800,000; and their investment was preserved in the company's later receivership. Bangor, the easternmost of the nation's ambitious seaports, contributed an equal amount to the Penobscot and Kennebec. Other acts were passed permitting aid to smaller lines.

During the fifties, also, Poor was attempting to secure private and governmental support for his "plan for shortening the time of passage between New York and London" by "extending a line of railway across the State of Maine and the Provinces of New Brunswick and Nova Scotia to the nearest available point of North America to Ireland." [23] This was a second bold project for a road crossing national lines. For the present, however, the "First International Railway" was achievement enough. In backing it with an investment of $2,000,000, the city of Portland had given one of the most notable examples of municipal aid.

West of the Appalachian Barrier, the most important internal improvements were intended to complement and extend, rather than compete with, the Erie Canal and its rivals. Connection between the Great Lakes and the Atlantic stimulated the idea of further connection with the Mississippi Valley, and lake ports from Pennsylvania to Wisconsin offered themselves as starting points for canals to the Ohio or the Mississippi. In the railroad area, completion of the trunk-line railroads to Lake Erie or the Ohio River—and the Grand Trunk to "Huron's waters"—led in the same way to eager competition in the building of connecting lines to the westward. In each of the states of the Old Northwest, the public authorities took part in the creation of improvements intended to serve these purposes and to meet the other needs of a rapidly growing frontier area.

Ohio's program was an early and substantial one. Routes for canals across the state from lake to river were surveyed while the Erie Canal was under construction. In the year of its completion Ohio adopted a comprehensive public works measure; De Witt Clinton came over from New York to help Ohio's governor to start construction on the first canal on July 4, 1825. The act authorized borrowing on the credit of the state, provided for an annual canal tax to be applied to the servicing of the loans, and established—as in New York—both a Board of Canal Commissioners and a Board of Commissioners of the Canal Fund. The first works authorized were the Ohio Canal, to extend across the entire state from Cleveland in the northeast by way of Columbus in the center to Portsmouth on the Ohio River, and the Miami Canal, to be built northward from Cincinnati as far as Dayton, with the idea of ultimate extension to Lake Erie. The sixty-seven miles of the Miami Canal were completed in 1829, and the full length of the Ohio Canal was open for traffic in 1833, giving the state a total of 400 miles of canal at a cost of something over $5,000,000.

This early accomplishment promptly inspired extension and emulation, in part because of the substantial traffic attracted to the canals and in part because of the claims of the regions which did not share their advantages. It was, said the governor, incomprehensible that the wants of these portions of the state "should have been wholly overlooked in our system of improvements";[24] and the legislature proceeded in 1836 to correct the oversight by authorizing six additional undertakings—the extension of the Miami Canal to Toledo, the Wabash and Erie, Walholding, and Hocking Valley canals, the improvement of Muskingum River, and the Western Reserve and Maumee Road. Of these improvements, the first was to provide a north-south connection from Cincinnati through the western part of the state; the second represented Ohio's share of the enterprise of joining Lake Erie with the Wabash River in Indiana; the other canals were laterals to connect with the main Ohio Canal. For the first two, the federal government had also contributed support. The state of Ohio received a land grant of 438,301 acres for the Miami Extension, and 292,224 acres of Ohio land granted to the state of Indiana for the Wabash and Erie were transferred to Ohio when the latter state agreed to build the portion of the canal that fell within its own borders.[25]

During the same period, enthusiasm for internal improvements added the method of mixed enterprise to that of public works. Early in 1837 the legislature passed a general law requiring the state to provide its aid to any railroad, turnpike, canal, or slack-water navigation company meeting certain standard conditions. In the case of railroads, the state's support was to take the form of a loan of one third of the capital needed for construction. In the case of turnpikes, the state was to subscribe for half the stock. In the case of canals and navigation companies, it was to take a third of the stock. Under this law, twenty-six turnpike companies, two canals, and three railroads received nearly $3,000,000 in aid from the state.

This lavish and optimistic policy, however, was shaken by the financial crisis of 1837 and did not survive the long depression that began in 1839. Enthusiasm for new projects evaporated when it became clear that the state could no longer borrow at a premium and when it became difficult or impossible to sell new bonds even at heavy discounts. Yet this revulsion of feeling affected the two parts of the program in somewhat different ways. The mixed enterprise measure—called the "Plunder Law" by its opponents—was repealed in 1840. With the exception of the loan to the Little Miami Railroad, which became a good investment after conversion into stock,[26] the state had little to show for its venture into mixed enterprise. With respect to the public works, no new measures were added, but the entire ambitious program of 1836 was carried through to completion. The Board of Public Works convinced the legislature that abandonment of works in process would be bad economy, and the Canal Fund Commissioners managed to scrape together the necessary funds —by selling bonds at discounts that reached a maximum of forty percent, by obtaining temporary assistance from local banks, and for several years by paying contractors in postdated checks. As a result, Ohio, unlike either New York or Pennsylvania, was able to continue its program "without interruption, even at the nadir of the depression,"[27] as Harvey Segal has pointed out, until the last of the canals was opened in 1845.

When the system was completed, the state owned 731 miles of canal, 91 miles of slack-water navigation, and a 31-mile turnpike. These had been built at a cost of a little less than $16,000,000. Canal revenues exceeded current expenditures from 1827 through 1855; and the net return exceeded $500,000 in 1847 and totaled more than $6,000,000 for the entire period to 1861. These sums fell far short of meeting interest charges. The showing would have been better if the lateral canals had not been built, and it is clear that the Board members made no serious attempt to justify their findings that these would yield "a revenue to the State, sufficient to meet the interest on the cost of the improvement."[28] On the other hand, both contemporary and later judgments agree that the canals, and particularly the earlier ones, had a substantial effect on the economic development of large areas of the state. Early hopes of a large through movement between lake and river proved illusory; but the canals carried a heavy traffic, particularly of the products of Ohio farms and mines, and continued to do so until railroad competition made them largely obsolete in the fifties.

With the credit and resources of the state government strained by its effort on the canals, the promoters of railroads turned to local authorities for assistance. For a time the response was enthusiastic. More than a hundred special laws, as well as a general measure passed in 1846, were enacted to authorize local aid. Almost every one of the railroads projected and being built between 1836 and 1850 asked and received subscriptions or loans from the counties and cities which it was to serve. In this movement, the city of Cincinnati played a leading role, aiding the Little Miami with a subscription of $200,000 and a loan of $100,000, lending $100,000 to a road to the eastward, and subscribing $200,000 to the Ohio and Mississippi in the hope of a St. Louis connection. The city also obtained authorization to subscribe $200,000 to South Carolina's grand project, the Cincinnati, Louisville and Charleston, and would have contributed $1,000,000 more to the Ohio and Mississippi in 1850 if the Illinois legislature had been willing to grant passage across the state. In all, the totals of local aid to railroads were about $8,000,000.[29]

In Ohio, as in other states, revulsion followed the early enthusiasm. The constitution adopted in 1851 prohibited both local and state aid to private companies. The rapid extension of railroads during the rest of the decade was accomplished with private capital, of which a considerable portion was provided by the four trunk lines, each of which established through connections across Ohio to the West. The state's interests in improvement corporations were disposed of to private interests, in most cases for very small return; and in 1861, the state canals were leased to a private company for a rental of a little more than $20,000 a year. These actions were "part of a general movement . . . of the alienation of state property and functions." [30] Ohio had been one of the major improvement states, with public investment amounting to some $27,000,000, and its early public works had made a substantial contribution to development. By the end of the period, however, with its program of public construction long since ended and with mixed enterprise prohibited, it stood as one of the chief examples of the revulsion of feeling against governmental promotion of internal improvements.

In three other states of the Old Northwest, the history of internal improvements resembled in timing and in results the more hectic and the less successful elements of the Ohio experience. In either 1836 or 1837, the years of the great expansion of the Ohio program, Indiana, Illinois, and Michigan each launched a highly ambitious program of internal improvements. Each of the three promptly encountered financial difficulties during the period of crisis and depression; each abandoned most of its unfinished program in 1840 or shortly after; and each went on to enact constitutional provisions against further state action. In two of the states, Indiana and Illinois, local governments assumed the functions of assistance given up by the states. Each of these two states had a lake-to-river project, like those of Ohio; the planning of these canals—and in Indiana the beginning of construction—antedated the more general programs. With the partial exception of these two undertakings, neither they nor Michigan could show anything to match the steady progress of Ohio's earlier canal program or its persistence in carrying its public works to completion during the depression.

Indiana's early hopes of capturing the trade between the Great Lakes and the Mississippi were based on the development of the old French and Indian trade route that made use of the Wabash River, which flowed across most of the state from the northeast to the Ohio, and the Maumee River, which entered Lake Erie at what is now Toledo. George Washington was one of the early proponents of a canal along this route, and in 1818 De Witt Clinton welcomed the proposal as a natural extension of the Erie Canal.[31] Federal support for the Wabash and Erie Canal, in the form of a land grant, was given in 1827. Construction as a state work began in 1832 but proceeded slowly and with many interruptions. By 1843, it had reached Lafayette in the upper part of the Wabash Valley; and the opening of the eighty-eight mile portion from the state line to Toledo, which the state of Ohio had agreed to build in exchange for its part of the lands granted by the federal government, provided the connection with Lake Erie. But Indiana did not succeed in carrying the canal to Terre Haute, midway down the Wabash, until 1848. Moreover, extension of the canal to Evansville on the Ohio River was not opened until 1853.

The completed canal, of more than four hundred and fifty miles, was the longest in the country. Its financial and economic results, however, were by no means commensurate. Construction costs amounted to almost $6,500,000, of which less than $3,000,-

000 came from the proceeds of the land sales; and net operating revenues over the whole life of the improvement were only about $1,300,000. Construction of the earlier parts of the canal had a substantial effect in reversing the direction of Indiana's trade from the Mississippi to the East and in promoting the settlement of the area. Even on the northern sections, however, tolls reached their peak as early as 1852, and the extension below Terre Haute never carried a heavy traffic and was abandoned in 1862.

A part of the difficulty with the Wabash and Erie arose in the entanglement of this enterprise with a more grandiose general program. Early in 1836, after the question had been the leading issue in the state elections, the legislature adopted what was referred to as the "Mammoth Internal Improvement Bill." This created a Board of Internal Improvements and an Internal Improvements Fund and authorized borrowing to the amount of $10,500,000 for completion and extension of the Wabash and Erie ($1,800,000), for a Central Canal ($3,500,000), for the Whitewater Canal in the southeastern part of the state ($1,400,000), for a highway paralleling the Ohio River ($1,350,000), for a railroad from Madison on the Ohio to Lafayette on the Wabash ($1,300,000), and for other smaller projects. The short history of the program was marked by attempts to start all the various projects at once, by corruption and inefficiency in organization, and by prompt financial disaster. Since no tax resources had been placed behind the Improvements Fund, interest charges were met out of borrowed funds. When no further loans could be secured, under the depressed conditions of 1839, the Board called a halt to all construction. More than $3,000,000 was lost when the Morris Canal and Bank Company and other financial agents, which had received state bonds on credit, failed in the depression. By 1841 the state was in default, with a debt of over $13,000,000, of which nearly $9,500,000 had been incurred for internal improvements. Against this, its improvement assets consisted of 200 miles of canal in use, yielding $5,000 yearly, two railroads yielding $26,000, and fragments of other canals.

Under these circumstances, the legislature in 1842 abandoned the policy of public works. It authorized the organization of private companies to own, complete, and operate all unfinished state works except the Wabash and Erie. Under these provisions the Whitewater Canal was completed in 1846 with aid from the city of Cincinnati. The Madison and Indianapolis Railroad, in which the state retained a stock interest, obtained local subscriptions, largely paid in land and produce, and managed to borrow eastern capital. It was finished in 1847 and for a time became a highly profitable road.[32] As a part of the state's settlement with its creditors, control over the Wabash and Erie was turned over in 1847 to trustees for the bondholders, and it was under their auspices that it was completed to the Ohio River. In 1851, with the public works program already abandoned, the revulsion against an improvements policy was further expressed by the adoption of constitutional provisions against state loans or subscriptions to private companies.

Yet in Indiana, as in a number of other states, public aid for improved means of transport was not completely checked by the rejection of state action. Particularly after 1845, the promoters of a large number of railroad projects appealed for aid to the interests of competing cities and counties. Sometimes these pleas were rejected—Indianapolis's Marion County, for example, voted against a subscription to the Madison road—and the sums involved were not so great as they were to be in later decades. Yet before the Civil War, railroad aid was given by some sixteen counties and ten cities, to an amount that may be estimated as about $1,750,000.[33] This local aid made a significant contribution to the rapid development which gave to the state more than 2,100 miles of railroad in 1860 and which had already fulfilled the ambition of an early governor to make Indianapolis the "Radiating Rail Road City." [34]

Illinois possessed the advantage of the best and shortest portage between the Lakes and the river system. From Chicago to a tributary of the Illinois River, the connection was so nearly a natural one that small boats occasionally made the passage in high water. The project of the Illinois and Michigan Canal was included in Calhoun's proposed system of national internal improvements, though not in Gallatin's, and was eagerly pressed by the state's representatives. In 1827, Congress granted aid for its construction by donating to the state of Illinois the alternate sections of land for a distance of five miles on each side of the proposed canal. The state appointed Canal Commissioners to carry on the work, but they were unable to sell a substantial amount of bonds on the basis of the land alone and had difficulty in raising money when the state's credit was also pledged to the venture. Actual construction began in 1836, but soon encountered the obstacles of the crisis of 1837 and the subsequent financial difficulties of the state. Though a number of the contractors accepted payment in bonds and interest-bearing checks, work was halted in 1841 and was not resumed until 1845. At this time, plans were altered to provide for a so-called "shallow-cut" canal instead of the more costly lake level project; and control was given to a board of Trustees, two representing the bondholders and one the state. Under the new management, and with additional funds provided by the bondholders, the work was completed and the canal opened in 1848.

Outlays on construction amounted to about $6,500,000, though the total costs were made greater by the accruals of interest during the period in which work was suspended. In the end, the proceeds from the sales of land, including Chicago lots, and the revenue from the canal itself, proved ample for the discharge of the indebtedness, so that the improvement was constructed without net cost to the state or loss to the investors. Its economic impact was immediate and important, and its construction was one of the major reasons for the phenomenal growth of Chicago as a transportation center.[35]

By contrast, the remainder of the state's program of internal improvements was an almost wholly unrelieved failure. Early proposals for mixed enterprise were rejected, and a comprehensive program of state works was inaugurated in 1837. The southern portions of the state objected to the concentration of expenditure in the north, and every section developed projects for improved transportation. After the bill prepared in committee had been added to on the floor of each house, it had indeed something for everyone: improvements for five rivers; three east-west railroads across the state, with various branches; and a great central railroad to extend from the northwestern corner to the southern tip of the state. In addition, presumably on grounds of what Gallatin had once described as "policy no less than justice," the act appropriated $200,000 for improvements in counties which did not share in the specific appropriations. The total expenditure authorized was $10,500,000, and the legislature prescribed that work should commence simultaneously on all the projects. Adoption of the measure was received with enthusiasm. The next legislature added an additional $1,000,000 to the program and a tax measure to give it partial support.[36]

Disillusion followed swiftly, however, when it became clear that no such sums could be either raised or serviced by a young commonwealth in a depression period. The major part of the program was abandoned in 1840, and in the following year the state defaulted on its bond interest. By this time some $5,600,000 had been borrowed under the act, besides the obligations incurred for the Illinois and Michigan Canal. Construction continued only to complete a part of one railroad, which was later sold for a fraction of its cost.[37] The rest of the work consisted only of fragments, and the heavily burdened frontier state remained in default until 1847. A new constitution, adopted in the following year, limited further state borrowing to $50,000 and prohibited aid to private companies.

Even after this experience, however, neither the legislators nor the people of Illinois were willing to rely entirely on private enterprise to bring them the railroads which they desired. In

1849, the legislature passed a general law allowing each town and county to buy or guarantee railroad bonds to the amount of $100,000, and some sixty special laws authorizing local aid were enacted during the eighteen fifties. Among the larger municipal subscriptions were those of Peoria, Alton, and Quincy. In the two latter cases, as in smaller Rock Island, the spur to public investment was the ambition to provide a major crossing place for trans-Mississippi railroads. The more than $4,000,000 of local government investment provided substantial aid to the railroad building of the decade.[38] Yet Chicago, with its advantages of position, was able to become the nation's greatest railroad center without contributing a cent of city funds to railroad development; and the state's longest railroad, the Illinois Central, was a product of federal rather than state or local assistance.[39]

Michigan became a state in 1835 at the height of early enthusiasm for railroad building, and its first constitution contained the injunction: "That internal improvements shall be encouraged by the government of the state." In his messages to the first two legislatures, the governor recommended meeting this obligation by the method of mixed enterprise in order to "combine the resources of the state with the interest and enterprise of the individual." Instead, the legislature adopted in 1837 a comprehensive program of state works. The sum of $550,000 was appropriated to begin work on three railroads across the state—the Central (of which a few miles had been built by a private company), the Southern, and the Northern. To finance the program the governor was instructed to negotiate a loan of $5,000,000. Difficulties soon appeared with respect to construction, with controversies over the choice of routes, with costs exceeding early estimates, and with charges of mismanagement and corruption; but the difficulties with respect to borrowing were greater and more decisive. After the Barings and other financiers had refused to take the bonds, the problem appeared to be solved by the sale of the entire issue to the Bank of the United States, operating under its Pennsylvania charter, and to the associated Morris Canal and Banking Company. The two companies, however, went into bankruptcy in 1839, after hypothecating the bonds; and the state received only about half the proceeds.

Opponents of internal improvements were victorious in the elections of 1840, and the program was abandoned, except for continuation of work at a reduced rate on the Central and Southern railroads. In 1842 the state repudiated the parts of the bond issue on which it had received no payment, and in 1846 it decided to sell the two railroads in order to meet part of the remaining debt. The Michigan Central, which was in operation for the 145 miles west from Detroit to Kalamazoo and was earning a substantial revenue, was sold for $2,000,000. The price was considerably below its cost to the state and proved an excellent bargain for John Murray Forbes and his associates who provided the capital and enterprise to extend it to Chicago and make it one of the nation's major railroads. The Michigan Southern, which had been constructed for about seventy-five miles westward from Monroe, had made less progress. After a brief period of operation as a municipal enterprise of the city of Monroe, it also was sold to private interests. The state received $500,000, less than half of what it had spent on the road, and in this case the purchasers altered its location before they entered into competition with the Central for a Chicago connection.[40]

After these transactions, Michigan's net outlay on construction, aside from interest and the proceeds of some 300,000 acres of land, had been under $2,000,000—a sum small by comparison with the programs of many states but large for the new commonwealth. The state's venture in internal improvements was so universally regarded as a failure that prohibitions against both public works and mixed enterprise were voted almost without discussion for inclusion in the constitution of 1850. A loophole was left, however, for the transmission of aid that might be offered by the federal government; and in this way Michigan obtained an improvement of lasting importance. In 1852 Congress granted to the state 750,000 acres of mineral and timber land to make possible the building of a canal at Sault Ste. Marie to connect the waters of Lake Superior with those of the lower lakes. A group of capitalists from the East and from Michigan itself obtained the contract from the state government, and the work was completed in 1855. Though its cost of $1,000,000 was double the original estimate, the principal investors found it the "handsome operation" they had anticipated,[41] since the grant included much of the richest copper land of the Upper Peninsula. The state of Michigan was left in possession of a canal, on which it levied tolls at a rate intended only to cover operating expenses and on which traffic in timber and wheat and, above all, iron and copper ore continued to increase. The present Soo Canal, as is often noted, carries a greater tonnage than any other in the world; and an orator at the semicentennial celebration, with an exaggeration appropriate to the occasion, claimed that the mile-long waterway, as "known by its results," was "the biggest thing on earth."[42]

In spite of the value of the canals ultimately constructed at Chicago and Sault Ste. Marie, it is clear that each of these three states had undertaken a program of internal improvement far too elaborate for its means. As Governor Carlin said for Illinois, they had been "allured from the path of wisdom and economy by the seductive spirit of speculation" in the inflationary boom of the middle thirties.[43] Nor is there doubt that contemporary criticism of their planning and administration was largely justified. Though at least half of Michigan's railroad effort was spent on a well-chosen route, Indiana failed to give priority to the more promising parts of its program, and the Illinois administration was forbidden by law to do so. Yet it must be conceded that the three states were also most unfortunate in the time at which they began their operations. If they could have had several years of construction before entanglement in the toils of the depression, the results would have been less dismal and the revulsion perhaps less complete.

From the experience of these states, their neighbor, Wisconsin, drew a lesson of self-denial. It entered the union in 1848 with a constitution forbidding both state works and loans to private companies. Yet its inhabitants had shown an early interest in internal improvements, and they made extensive use of local government agencies for their support.

During the territoral period, Wisconsin had its own project for a lake-to-river connection, from Milwaukee to the Rock River, which was intended in part to divert from the Mississippi to Lake Michigan the production of the lead mines in the southwestern part of the area. In 1838, the federal government granted land in aid of the proposed canal and later increased the amount to a total of 500,000 acres. But neither the lands themselves, nor financial aid voted by the territorial government in 1840, were ever made available to the company. At the end of the decade, the company claimed the lands in aid of a railway rather than a canal, but the state decided to use the grant for educational purposes instead. In 1856, on the other hand, the legislature after long controversy divided between two railroad companies a large land grant made by the federal government. The first of these companies, which became a part of the Chicago and Northwestern, built toward the northeast and ultimately received nearly 550,000 acres. The second, which was to build to Lake Superior, was found to have won its grant by bribing the governor and members of the legislature; and the large acreage intended for it was later divided among several other companies.[44]

In the constitutional convention, there was a strong minority of delegates who argued that, with proper safeguards and restrictions, Wisconsin could construct a useful system of internal improvements without falling into the errors of its neighbors. In the year following the defeat of their proposal, the improvement advocates turned to the alternative method, and a law was passed authorizing the city of Milwaukee to aid the Milwaukee and Waukesha Railroad. During the decade of the fifties, nearly eighty measures were enacted authorizing local aid, and county, city and town bonds were issued to the amount of about $7,250,000.[45] This was a decade of rapid railway construction, in which nine hundred miles were built, and the network of lines that covered the southern part of the state by 1860 received a large part of their capital from this source. In the case of Milwaukee, the policy was pursued until, as the mayor could boast in 1857, "the different railways having Milwaukee as a terminus [were] mostly in that state of progress that they need no further aid from the city."[46] But many of the farming localities were less complacent about their railroad contributions, particularly to companies that collapsed in the crisis of 1857, and several municipalities repudiated their railroad bonds before the decade was over.[47]

To the west of Wisconsin lay the still newer state of Minnesota. Though it obtained no railroad mileage before the sixties, its history displayed a remarkable series of alternations in sentiment and legislation. During the territorial period, there was agitation for federal aid to railroads: the Minnesota and Northwestern Railroad Company was organized in 1854 in the hope of receiving a land grant, which was made but later revoked. When Minnesota entered the union in 1857, its first constitution followed the Wisconsin pattern, forbidding both public works and aid to private companies, except as a trustee of possible federal grants. In the following year, the position was reversed by constitutional amendment, and bond issues were authorized in aid of four railroads up to $1,250,000 each. The railroads were entitled to half the sum on the basis of grading alone, and bonds were issued to the amount of $2,275,000. But the companies had almost no private resources, and little work was actually done.

In 1860 there was a sharp revulsion against the improvements policy. The "swindling bonds" were repudiated, to be compromised some years later, and further bond issues were prohibited by constitutional amendment.

In the same tier of states, Iowa was older and more populous than Minnesota and lay on the direct route from Chicago to the West. Construction of 655 miles of railroad by 1860 was largely the result of these advantages, supplemented by substantial local assistance. Adopting its constitution in 1846 under the influence of the difficulties experienced by Illinois and other states, Iowa entered the union with strict limitations on the creation of state debt and a prohibition against state participation in mixed enterprise. Here as elsewhere, however, local governments came forward eagerly to extend the aid denied by the state. By 1858, when the practice was held unconstitutional by the courts, they had invested an amount which has been estimated as more than $7,000,000, though the proceeds to the companies, after deductions for discounts and commissions, were in many cases not much more than sixty percent of the face value.[48]

Of all the western programs of state and local aid, the largest was that of Missouri. Its timing, however, differed significantly from most of the others, although railroad building was enthusiastically advocated, in Missouri as elsewhere, in the middle thirties. A committee appointed by an 1836 railroad convention in St. Louis proposed a statewide system of improvement which would

> approximate the east, west and middle country . . . break down sectional animosities . . . increase the value of agricultural products, extend commerce and aid in the development of unexplored resources.[49]

Provisions permitting local aid were included in several railroad charters, and there was strong pressure in favor of state assistance as well. However, the crisis of 1837 intervened before further action was taken or construction begun. Thus Missouri, without public commitments, avoided the financial difficulties and the disillusion that beset so many improvement programs during the depression years.

When prosperity returned, the demand for improvements was renewed. The first local government aid, however, was not given until 1850, and the first state aid was given in the following year. By this time the desire for railroad development was intensified by the hope that Missouri, and particularly the city of St. Louis, might become the principal terminus of a projected railway to the Pacific. With this in mind, the governor recommended that the state's program be concentrated on aiding two main trunk lines to cross the state from east to west; and an act of 1851 authorized a loan of $2,000,000 to the Pacific Railroad, to extend from St. Louis to Kansas City, and of $1,500,000 to the Hannibal and St. Joseph, to cross the northern part of the state. These sums were soon found insufficient to secure actual construction, and the interests of other projects and other sections soon broke down the policy of concentration. Successive appropriations in 1852, 1855, and 1857 raised the total authorizations for state aid to almost $25,000,000, of which $23,701,000 was actually extended. Recipients of the new loans, in addition to the two roads originally designated, were the Southwest Branch, projected to take advantage of a federal land grant; a north-south road, the North Missouri; the St. Louis and Iron Mountain; and—of smaller amounts—the Platte County and Cairo and Fulton Railroads.

Each of these railroads also received support during the same period from local governments. The total local aid authorized through the year 1860 was over $9,000,000, and the amount actually extended was $8,335,000. The bulk of the amount—over $6,000,000—came from the city and county of St. Louis: they subscribed $2,200,000 to the Pacific Railroad and $700,000 to the Ohio and Mississippi, which gave the city its connection with the

Baltimore and Ohio and the East.[50] The remainder of the aid came in much smaller amounts from thirty-five other counties and four small towns. Subscriptions were usually paid in bonds, but the county contributions to the Cairo and Fulton were mainly in swamp lands taken at the valuation of a dollar an acre. Under most of the laws, the county courts were allowed to make subscriptions on their own authority, but they might—and, in fact, usually did—take a referendum of the voters "for information." The county stock was generally distributed to the property holders in order, as a railroad official explained, to "constitute the taxpayers stockholders in the Co. thus giving every taxpayer an interest in the road in proportion to the amt of his taxes" and "a voice in the election of directors and location of the road." [51] In a law affecting St. Louis County in 1857, the conclusion was drawn from this practice that only property owners should vote on the question of subscription. Whatever the method, public support for the program appears to have been general, and few appeals for aid were rejected, though in some cases the authorized subscriptions had to be declined because the railroads were not willing to meet the specified conditions of location.

If the purpose of mixed enterprise was to draw out private wealth for public ends, its success in Missouri was distinctly limited. By 1860 the public authorities had committed over $32,000,000 to railroad aid, though its value to the companies was lessened by the fact that state and local bonds had often to be disposed of at a discount. Even allowing for this, they seem to have provided about three quarters of the total amount spent on construction.[52] Almost all of the stock of the St. Louis and Iron Mountain was taken by the city and county of St. Louis; local governments subscribed about three quarters of the capital of the North Missouri; the Southwest Branch had almost no private subscriptions except from contractors. The notable exception was the Hannibal and St. Joseph which profited by federal, state, and modest county aid but also drew in a large amount of eastern capital, built more substantially than the other roads, and became a profitable part of the Burlington system.

By 1860 about eight hundred lines of railroad were in operation. The Hannibal and St. Joseph had been completed the year before, but the Pacific was still a hundred miles short of Kansas City, and the other roads remained mostly unfinished. Each of the state's seven railway companies—again with the exception of the Hannibal and St. Joseph—was in default on its bond interest.

With the coming of the crisis of 1857, the bonds of the state and counties became unsalable except at heavy discounts, and additions to the aid program were almost completely suspended. In 1858 an amendment to the constitution limited to $30,000,000 the total of state debt that could be contracted "for the prosecution of internal improvements" or for any other than emergency purposes, and laws of 1860 and 1861 tightened the conditions under which county aid could be given. Thus at the end of the period Missouri had begun to show at least partial evidence of the revulsion against internal improvements which had occurred in a number of other states under similar circumstances two decades before.

In the part of the country south of the Ohio River, the state of Kentucky showed an early interest in internal improvements but devoted relatively little state money to their construction. Its first important project was a canal around the Falls of the Ohio at Louisville. Since these formed the one serious obstacle to navigation in the whole length of the river, the Kentucky Legislature had reason for asserting, as it did in 1805, that the proposed improvement was of importance to the whole "Western country . . . as well as to the United States generally." A law passed in the same year provided that blocks of stock in the projected company should be reserved for subscription by Kentucky itself, by other interested states, and by the federal government.[53]

Twenty years later, there was strong support for the construction of the canal as a state enterprise; but the legislature defeated the proposal by a close vote and thus, according to the Governor, "threw away a source of revenue as lasting as that noble river." Instead a private company was chartered in the hope "to tempt the employment of foreign means in the undertaking"; and the canal was built by individual subscriptions supplemented by the national government.[54] In the year that the canal opened, 1830, another and more famous Kentuckian plea for federal assistance became the subject of the Maysville Road veto of Andrew Jackson.

Kentucky's first railroad, which was also one of the first built west of the Appalachians, provided an early instance of municipal aid. In 1831, the city of Lexington subscribed $25,000 for stock in the Lexington and Frankfort, which was completed in 1835 after it had replaced its original horses and mules with a locomotive built by a Lexington mechanic. Extension of the line to Louisville proved to be more of a problem. The Louisville and Frankfort, which received state aid during the same decade, was taken over, put into condition and operated by the state government in the early forties. This was the state's principal activity in internal improvements, and further participation was precluded by the adoption in 1850 of a constitutional prohibition against the loan of the state's credit.

The state's total expenditure on railroad development was less than $750,000, and in 1857 an organ of the industry declared that, "Of all the States, Kentucky has made the poorest show in the matter of internal improvements."[55] Yet the totals of county aid (over $7,000,000) and of municipal aid (nearly $9,000,000) were both substantial;[56] and they played a considerable role in an active program of construction during the fifties, which brought Kentucky's railroad mileage to 534. In this the leading role was taken by the ambitious city of Louisville, which contributed no less than $5,000,000. Of this about $800,000 represented an early investment in its eastern connection, the Louisville and Frankfort;

and $200,000 was its subscription to the Jeffersonville Railroad, which was building northward from a point across the river in the effort "to draw away from Cincinnati the trade of the fertile state of Indiana." [57] Much more important was the support given to the Louisville and Nashville, which was to become one of the principal railroads of the South. Its main line connecting the two cities was completed in 1859 at a cost of a little more than $7,000,000. Of this a part had come from Kentucky counties and another part from public aid on the Tennessee side, but $2,000,000 was contributed by the City Council of Louisville.[58]

By contrast with Kentucky, the neighboring state of Tennessee gave extraordinarily full support to internal improvements by action at every level of government. Aid was extended mainly during the late eighteen-thirties and the decade of the fifties. In the first period, which has been described in a book entitled *Sectionalism and Internal Improvements*,[59] the principal role was taken by the state, and the largely abortive efforts were followed by a sharp revulsion against public aid. In the second period, when renewed state action was widely supported by county and municipal aid, Tennessee constructed in a single decade and largely with public funds one of the three largest railway networks in the South.

Even before the Maysville Road veto, legislators in Andrew Jackson's own state drew from his national policy the moral of state responsibility for development. As the House Committee on Internal Improvements declared in January, 1830, "The States, secured from encroachments, are thrown back on themselves." [60] In the same year, a Board of Internal Improvements was established and $150,000 appropriated for river improvement. With the $60,000 allotted to the eastern portion of the state, enough was done on the Tennessee River to permit the first steamboat to reach Knoxville. Meanwhile the enthusiasm for railroads, which a doubting Tennessean described as being "rode 'hobby-horsically' throughout the United States," [61] had already reached his own state. In 1831 and 1832, an "Association of Gentlemen" in the mountain village of Rogersville published the twenty-six fortnightly issues of *The Railroad Advocate*, the first journal of the American railway industry. As this suggests, the railroad interest was particularly strong in East Tennessee; but there was also active agitation at the other end of the state: leading citizens of Memphis formed ambitious projects of railroad connections with Charleston in the southeast, by way of the Muscle Shoals, and to Baltimore by means of a road extending northeastward across the full length of the state. The more prosperous areas of central Tennessee, however, and particularly the city of Nashville, were on the whole hostile to railroad aid and looked instead to the improvement of their immediate area by turnpikes.

The constitutional convention of 1834 adopted a provision laying upon the state the responsibility of encouraging internal improvement. Positive action, however, encountered formidable obstacles in the clash of local interests and the general shortage of local capital. An act of 1836 pledged the state to subscribe one third of the stock of any railroad or turnpike corporation after it had succeeded in raising the other two thirds by its own efforts, but the only companies that succeeded in meeting this condition were three turnpikes in Central Tennessee and the Memphis and Lagrange Railroad, which was organized to build the first link of the connection with the southeast. In 1838 a new measure was passed, mainly by the votes of legislators from East and West Tennessee, which raised the proportion of the state's subscription to one half of the company's capital. Total aid under the act was to be limited to $3,700,000—$1,300,000 in East Tennessee, $1,500,000 in Central Tennessee, and $900,000 in the Western District, with an additional $100,000 allocated to river improvement in each of the districts. Of the allotment to East Tennessee, half was to go to the great Charleston-Cincinnati project and half to the Hiwassee Railroad, which was to build toward the northeast from a junction with Georgia's state railroad.

This extensive program, like those of many other states, failed to survive the depression years. Central Tennessee obtained the completion of its network of turnpikes radiating from Nashville; but the Charleston dream was abandoned, Hiwassee construction was long delayed, and no railroads were brought into operation. The act was repealed in 1840, though with the proviso that the Memphis and Lagrange subscription be brought up to the fifty percent standard. Some further aid was given to this road later in the decade, but a governor summed up the prevailing sentiment with the admonition: "In the name of the honor of the State, *issue no more bonds*." [62]

Returning prosperity, however, and the continued lack of railroad communication brought about a sharp reversal of policy. A law of 1852 provided for state loans to railroads at a standard rate of $8,000 per mile, and in 1854 the figure was raised to $10,000. The problem of sectionalism was met by making this assistance available to any corporation, domestic or foreign, that could raise the necessary capital. Loans extended under this legislation brought the total of state aid to more than $17,000,000, and this was almost universally supplemented by the resources of the local governments. Municipalities, led by Memphis and Nashville, contributed nearly $3,500,000; and another $5,000,000 came from no less than seventy-seven counties. At the end of the decade, the Louisville and Nashville and its branches had reached Memphis and Nashville; both cities had connections with Charleston and Savannah; the successor to the Hiwassee had a through route from Chattanooga through Knoxville and into Virginia; and the state had 1,254 miles of railroad. To this burst of construction, public authorities had contributed nearly $26,000,000 of a total investment that has been estimated as about $40,000,000.[63]

In Alabama, the state did little more than pass on to particular companies the aid given by the federal government, but the city of Mobile provided a notable case of municipal railroad promotion. As in neighboring Georgia, early projects for a canal to connect the Tennessee with streams running directly to the Gulf of Mexico were never carried out. Later agitation for railroad aid met the obstacles of a state credit impaired by the failure of the State Bank and the opposition of a "veto governor" elected to office on the program of restricting the role of government to the performance of "few and simple" functions. Like other "public land" states, however, Alabama received from the federal government five percent of the proceeds of public land sold within its borders. Of these funds, $135,000 were spent on river improvement, some $400,000 were devoted, in the forties and fifties, to attempts to secure railroad connections crossing the state from east to west and from north to south; and, just on the eve of the Civil War, an additional $850,000 were distributed as loans to eight railroads. A loan of $400,000 to the Mobile and Ohio was one of the few cases of aid extended from the state's own resources. The total of state aid was a little over $2,000,000. In addition, the state received and transmitted to the companies some 3,000,000 acres of land donated by the federal government under the railroad land grant legislation of the eighteen-fifties.[64]

County aid in Alabama amounted to barely more than $500,000. Municipal authorities, however, contributed more than $3,000,000. Montgomery, Tuscaloosa, and Selma all invested substantial amounts; but more than half of the total came from Mobile, and of this the greater part was devoted to the Mobile and Ohio. In this project, launched in a public meeting in 1847, "the business men of a city of less than twenty thousand," as a local historian boasted, "undertook to build a railroad five hundred miles long."[65] The "grand conception" was to intercept and bring to Mobile the trade of the Ohio and Tennessee rivers, and perhaps the Mississippi, by building a railroad across country to a point near their junction.[66] To finance construction, the business men of Mobile had to enlist public aid from a variety of sources. The Mobile and Ohio, together with the Illinois Central with which it was to connect, became the beneficiaries of the first federal land grant under the law of 1850. Loans from the state of Mississippi were added to that from Alabama, and aid was also secured from Mississippi counties. Mobile itself made municipal subscriptions of $1,100,-000, after the city's property holders had agreed with only seven negative votes to a five-year tax for the purpose, under an 1851 law which permitted deductions for those who had already made private subscription and which provided for distribution of the city's stock to individual stockholders.[67] With all this aid, construction proceeded at a slow pace, and the road reached its destination at Columbus, Kentucky, only on the very eve of the Civil War. A substantial traffic in cotton had already developed, but the outbreak of hostilities deferred a test of the road's effectiveness in fulfilling Mobile's hope of "emptying the Ohio and the Tennessee into her basin." [68]

In Louisiana, a period of early aid was followed by a revulsion even more sharply marked than that of Tennessee; but a revival of state and local assistance in the fifties enabled New Orleans to outstrip Mobile in building a railroad to the North. Legislation in 1833 followed the conventional pattern of establishing a Board of Public Works, an Internal Improvements Fund, and a State Engineer. During the decade the Fund made railroad investments of about $1,450,000, mainly in the form of loans. Most of these sums were lost in defaults and foreclosures, and in 1845 Louisiana became one of the earliest states to forbid state subscriptions and loans to internal improvement companies.

The policy of abstention, however, did not survive the return of prosperity and the growing realization of New Orleans leaders that the supremacy of its great river was threatened by railroad competition.[69] A law of 1852 permitted "parish police juries and municipal corporations" to subscribe to railroad corporations, with the taxpayers becoming "individually owners" of the stock. In the following year, the constitution, which the *American Railroad Journal* had described as "antiquated," was "remodelled, so as to permit extension of aid to railroad projects." [70] Under its provisions, which permitted state subscriptions to the extent of one fifth of the paid-up capital of a corporation, the legislature appropriated nearly $2,000,000. Parish or county aid amounted to about $450,000, and the municipalities contributed nearly $5,000,000, almost all of which was from the city of New Orleans. The two principal enterprises bore the resounding titles of the New Orleans, Opelousas and Great Western and the New Orleans, Jackson and Great Northern. The former had the immediate object of serving the sugar regions and the larger ambition, toward which little progress was made before the war, of building to the Texas border and becoming "the main trunk" of a Pacific railroad. It received $1,500,000 from the city and $700,000 from the state, in addition to local aid from the sugar parishes and a federal land grant. Still larger sums—$2,000,000 from the city and nearly $900,000 from the state—went to the Great Northern, which was to build through the cotton districts of Mississippi to the same destination as the Mobile and Ohio. Its New Orleans proponents left no doubt that the project was largely inspired by the rival enterprise of their "little neighbor, Mobile." [71] With aid from the states of Mississippi and Tennessee, the "Jackson Road" and its northern connections completed their through route (which is now the southern division of the Illinois Central) in January, 1860. By a year's margin, New Orleans had won the race with its smaller rival.[72]

The greater part of Mississippi's railroad mileage in the antebellum period was supplied by the rival roads from Mobile and New Orleans, each of which provided outlets for cotton produced in areas that did not have the benefit of Mississippi River transportation. The Mobile and Ohio received a state loan of $200,000 and generous subscriptions from the eastern counties. The Jackson Road received a state subscription of $430,000 and a loan of $200,000, and its northern extension, the Mississippi Central, received a somewhat smaller loan. The two north-south lines were intersected by the Southern, a purely Mississippi enterprise built east from Vicksburg.

A much earlier project, the Mississippi Railroad starting at Natchez, had been launched in the eighteen-thirties, and received state aid in bonds and also by gift of stock in the state-owned Planters' Bank and by grant of banking privileges. Both bank and railroad, however, went into bankruptcy, and the enterprise was abandoned until after the Civil War. As in the case of Alabama, the state's credit was weakened by the banking failure, and most of the state's aid to railroads, which totaled a little more than $2,000,000, represented reallocation of funds received from the federal government. Municipal aid was $425,000. The substantial amount contributed by the counties, which was more than $1,600,000, reflected the fundamentally rural nature of the state's economy.[73]

Florida obtained about 400 miles of railroad before the Civil War, with two lines almost completed across the peninsula but none connecting with the railroads of other states. Support of the enterprises came almost entirely from public sources, including land grants from the federal government. In 1855, the state legislature adopted an act "to provide for and encourage a liberal system of Internal Improvements." The aid extended, as in Alabama and Mississippi, consisted largely of funds the state had received from the federal government. The total assistance from the state government approached $5,000,000, but more than half of the figure represents the estimated value of land given by the state. To this, local authorities added some $600,000.[74]

In frontier Arkansas, railroad projects during the eighteen-fifties were stimulated by the hope of providing the first link of a transcontinental road, but the only public aid actually given was $180,000 by the state. At the other end of the proposed route, California had already adopted legislation authorizing local aid; but the only railroad in operation by 1861 was built in the gold-bearing Sacramento Valley entirely by private means. Oregon had as yet no railroads.

The case of Texas is of somewhat greater interest not so much because of what was accomplished before the Civil War as because of the precedents established for later action. Here also there was lively discussion of routes to the Pacific, but most of the three hundred miles of railroad in operation by 1861 had the more modest purpose of connecting the interior of East Texas with the various Gulf ports. To one of these, the Buffalo Bayou, Brazos and Colorado, the city of Houston built a small connecting road, the Houston Tap. The purpose was a familiar one: the merchants of the city had begun "to see some of their trade going down the Bayou to Galveston instead of up the Bayou to Houston." [75] The financial results were somewhat more unusual: the city was able to sell its municipal enterprise for considerably more than the $132,000 it had invested.

There were several other cases in which local government aid was given in more conventional forms, but the principal encouragement to internal improvements was given by the state. During the first decade of Texas statehood, some of its principal political leaders, convinced that railroads were "the only Key that Texas has to unlock her casket of costly gems," [76] advocated the method of state enterprise. In 1855, Governor E. M. Pease put before the legislature a comprehensive plan for a coordinated system of rail and water transportation, to be built by the state and financed by borrowing on the security of the public lands which Texas had retained when it joined the Union. The legislature had, however, already adopted an alternative plan for applying this resource to the promotion of internal improvements. In 1852 two railroads were offered state land at the rate of eight sections a mile. In 1854 the principle was applied to all railroads, and the rate of subsidy was doubled. The governor of a neighboring state declared that Texas was acting "with a wisdom worthy her good fortune" in retaining control of these lands by bestowing them "bountifully for the purposes of internal improvement" and on terms more generous than those of the federal land grants of 1850.[77] Under the frontier conditions of the time, however, even this bounty appeared insufficient to encourage railroad investment; and in 1856 the state added further aid in the form of loans at the rate of $6,000 per mile. By 1861, state bonds to the amount of $1,800,000 had been issued to the railroads, and they had fulfilled the conditions for receipt of some 5,000,000 acres, though most of the land had not yet been patented.[78] The outbreak of war interrupted this program in its early stages, but the policies adopted foreshadowed later developments both in the state and in the nation.

Public policy toward internal improvements varied greatly, as these Chapters 3 and 4 have shown, among the thirty-three states that were members of the Union in 1860. Though the figures that have been presented differ in accuracy and completeness, and provide only an approximate basis for comparison, they show that the total effort was a great one and that it was very unequally distributed among the various commonwealths. During the period covered, state and local governments, in addition to granting aid in land and materials or in various privileges, appear to have made financial commitments for improvements amounting, at least at face value, to well over $400,000,000. Of this total almost sixty percent came from the seven states directly facing the Appalachian Barrier. Public expenditure approached $70,000,000 in New York and exceeded $50,000,000 in Virginia. The amounts spent in the two western states of Missouri and Tennessee are equally impressive if they are related to wealth and population. At the other extreme, there were several states in which no public funds at all were devoted to internal improvements.

The various programs differed also in the parts played by the different levels of government, in the emphasis placed on public works as against mixed enterprise, and in the nature of the improvements constructed. Though local governments appear to have contributed only about a quarter of the total, Baltimore's investment exceeded that of many state governments; and the cities of Portland, Louisville, and Mobile each spent more than its respective state. The three massive public works programs of New York, Pennsylvania, and Ohio, devoted entirely or principally to canal building, account for about a quarter of the total outlay. As the major interest turned from canals to railroads, much less reliance was placed on public construction; and most of the remaining expenditure took the form of subscriptions or aid to mixed or private corporations. The sequence of projects, as would be expected, reflected technological changes. Early expenditures in turnpikes and river improvements were soon overshadowed by the major outlay on canals, and this in turn was surpassed by public expenditure on railroad development, which for the period as a whole represented some sixty percent of the total.

There were also wide differences in the means chosen for the selection and administration of projects, as well as in their financial results and their effectiveness in promoting development. In some cases effort was concentrated on a single project of strategic importance to the economic development of the state as a whole. New York's building of the original Erie Canal and Georgia's construction of the Western and Atlantic Railroad were the most notable examples, and this was the policy for which Pennsylvania's "Mainliners" contested in vain. More frequently state programs included a variety of projects, often quite uneven in their economic importance, and reflecting the balance of political forces representing different parts of the state. In order to minimize such logrolling and to place responsibility on individuals and localities, several states, including Virginia in 1816 and Tennessee and Texas nearly forty years later, adopted the policy of aiding on equal terms all improvement projects within their borders.

Equally striking were the variations in the position of the various programs at the end of the period. Accounts of the history of internal improvements have stressed "a revulsion of sentiment" in the early eighteen-forties which is said to have "turned all this business over" from state and local governments to individuals and corporations.[79] Certainly the record of those years shows the abandonment of a number of major programs, either because the undertakings had been completed (as in the case of the Ohio canals), or more often (as in Michigan, Indiana and Illinois) because newly undertaken projects had come to grief in the depression years following 1837 and 1839. Yet in a number of cases, the revulsion against state action was followed by extensive reliance on local government aid. Moreover, certain of the programs—those of Massachusetts, Virginia and the city of Baltimore, as well as some in the western states—were at their greatest height in the decade of the fifties.

The amount of aid extended by almost every improvement state, and indeed that of each of several ambitious cities, exceeded the total amount devoted to internal improvements by the federal government during the entire Era of National Projects. Even on the routes which Gallatin had marked for federal action because of their national importance, states and cities did the greater part of the work. Of the four "necks" of land to be cut for the intercoastal waterway, the federal government had, to be sure, joined with state and private subscribers in completing two of the projects; a third had been constructed by private enterprise; and a fourth remained for private construction and federal purchase in the twentieth century. But on the much more important problem of the Appalachian Barrier, the federal contribution was limited to the National Road and an early subscription to the Chesapeake and Ohio Canal Company. Of the improvements that made such effective use of the Mohawk-Hudson opening, the Erie Canal was a state work and the New York Central a product of private enterprise with some municipal aid. On the Allegheny route, the first major improvement was also a state enterprise but it was superseded by the Pennsylvania Railroad which was built by a mixed-enterprise company. Successful exploitation of the Potomac route was by a corporation financed in the same way, namely, the Baltimore and Ohio Railroad; and the still unfinished attempt to complete a major connection to the West by way of the James and

Kanawha Rivers had engaged the efforts of both mixed and state enterprise.[80] The principal achievements of internal improvements had been the work not of the national government, as Gallatin and John Quincy Adams had wished, but of what the latter called "State legislature and private adventure." [81] Yet the federal government had begun to reenter the field in the decade before the Civil War, as has been indicated in the references to land grants to railroads during that period; and its contributions were soon to be greatly increased. It is necessary therefore to return to an examination of national debates and decisions.

FEDERAL DEBATE AND DECISION

1850-1872

It was a national work, originating in national necessities, and requiring national assistance. . . . The act itself was an experiment. It must be considered in the nature of a proposal to enterprising men to engage in the work.

United States Supreme Court,
in an 1875 decision on the
Pacific Railroad Acts

5

AN ERA OF NATIONAL SUBSIDY

A new era of national action began in 1850 with an extensive grant of federal lands to aid in the construction of a system of railroads from Mobile, Alabama, to northern Illinois. There had, of course, been federal action in an earlier period, as Chapter 2 has indicated. Advocates of the new measure reminded Congress of the encouragement given to "Old Colonel Zane" back in 1796,[1] and they could cite as a more direct precedent the grants to Illinois, Indiana, and Ohio in 1827 and 1828 which had introduced the practice of giving alternate sections of public land along the route of the proposed improvement. Nor had federal activity in the field ceased completely after the Maysville Road veto. Appropriations were made—though after Jackson's administration very sparingly—for river and harbor improvements. The National Road was pushed slowly on into Illinois. The federal government retained its stock in the Louisville and Portland Canal, which became a highly profitable undertaking; but the government's unwillingness to assume positive responsibility was illustrated by its acquiescence in the arrangement by which the profits were used to buy out the private shares and by its failure to take over full control even when the process had been completed.[2] Certain concessions and privileges were granted to the railroads. From 1830 to 1841, after an agitation begun on behalf of the Baltimore and Ohio, they were given either full or partial rebates on the duties on iron imported for rails. Moreover, the Tallahassee Railroad and three other companies in the Gulf states received free right of way through federal lands and the use of stone and timber from them.

More characteristic were forms of federal aid applied to internal improvements indirectly, through the agency of the several states and at their discretion. For each new state formed from the public domain between 1802 and 1820, and again for Iowa in 1841, a fraction of the proceeds of federal lands sold within its borders was allotted to the state government for road building or other improvements. In 1836 additional sums were made available to Alabama and Mississippi from the proceeds of the sale of the lands of the Chickasaw Indians. The federal government also continued to make grants of the land itself for internal improvement; but policy shifted from the national selection of particular projects, such as the Illinois and Michigan Canal, to a general distribution to the states for use for their own purposes. This latter tendency triumphed in the act of 1841, which brought up to a total of 500,000 acres the improvements grant for each of the public land states and established the policy of distributing the same amount to each new state on its admission.[3] Sometimes, as has been shown, the grants were not in fact used for the purposes intended. On the other hand, a large number of states applied to their internal improvement programs the substantial sums received from the distribution of the surplus federal revenue in 1837.

The 1850 act was a sharp departure from these precedents and from the prevailing attitudes of the thirties and forties. In the first place, it represented federal aid on the grand scale. The grant was to consist of alternate sections of land in a belt six miles wide on each side of the railroad. Where parts of this land had already been sold, the companies could take compensatory acreage within nine miles of the track. The amount finally granted was about 3,750,000 acres. In the second place, it was the Congress that established the termini and laid down the general direction of the route, though the act followed established forms by granting the land to the states of Illinois, Mississippi, and Alabama for retransfer to the companies. Finally, the act was similar in character to a number of measures for aid to railroads, including one for the Illinois Central, which had been rejected during the two preceding decades.

How, then, was this change of policy effected? Part of the answer no doubt lies in the mounting pressure of public enthusiasm for railroads in a period of exceptionally rapid construction. Much is clearly due to the strategy of Senator Stephen A. Douglas and other leaders who combined the southern project of the Mobile and Ohio with the northern project of the Illinois Central so that their proposal could offer "a continuous railroad communication from the Gulf to the extreme limit of the State of Illinois." This not only served to attract votes from the sections immediately concerned but also made it possible for the advocates of the improvement to base their arguments on its importance as "a great national thoroughfare." A representative from Alabama declared that a work of such preeminently "national character" would serve "to perpetuate this glorious Confederacy"; a senator from Illinois expressed the hope that it would bind the North and South "together so effectually that the idea even of separation" would be unthinkable. In the same debate Senator William H. Seward of New York attributed the financial difficulties of the newer states to "the circumstance that it was devolved upon [their] governments . . . to make works of internal improvement" while the resources best "applicable to that object," the public lands, "belonged altogether to the Federal Government."[4] In this argument revulsion against the failure of state programs like those of Illinois took the form of a return to reliance on the agency of the national government.

The government's land resources figured in the constitutional as well as the practical aspects of the discussion. Defenders of the proposal derived part of their constitutional justification from the government's position as proprietor. Like any other landowner, they argued, it could dispose of part of its property to enhance the value of the part that remained. As evidence of this intention, the 1850 act, like earlier grants, provided that the alternate sections reserved to the government should not be sold at less than twice the minimum price. This provision, however, drew attack from those who were against raising the cost of land to actual settlers; and it failed to remove the "radical, abiding and constitutional objections" of a number of legislators, particularly from the southeastern states, to any federal program for internal improvements. "Where," they asked, "is the power in this Government to make a donation to A in a manner that presses B into paying double price?" Observing sadly that constitutional scruples seemed greatly to depend "upon local patriotism" and to diminish as men moved toward the frontier, the opponents of the measure went down to defeat before the votes of the western states and their allies.[5]

The promoters of the Illinois Central counted on bonds secured by land as their major source of funds for construction, and their desire to avoid areas in which the land was already taken up was a factor in the choice of a route through relatively thinly settled parts of the state. The *American Railroad Journal* expressed doubt of the soundness of an enterprise in which "commercial considerations" were thus "sacrificed for the purpose of obtaining a large quantity of land." In the event, however, the combination of the native richness of the prairie along the route and the vigorous colonization work of the Illinois Central gave to the enterprise a strong developmental impact. Though the *Journal* considered the Mobile and Ohio as by contrast a work of "inherent strength" that "would have been built without adventitious aid,"[6] the seven hundred miles of the Illinois Central were in fact completed in 1856 well before the five hundred miles of the southern road.

Meanwhile the federal government was already adding further aid to railroad development. An act of 1850 turned over unsold swamp and overflowed lands to the states within which they lay, and in some cases these so-called "refuse lands" were later disposed of for improvement purposes. An act of 1852 granted to railroads generally the privilege of right-of-way and the use of material on public lands. In 1852 and 1853, Missouri and Arkansas received land grants for five railroads; and in 1856 and 1857, at the height of the railroad boom, the congressional barriers were lowered for a flood of measures intended to aid no less than thirty-five additional railway projects. These two sets of grants went to ten of the western and Gulf states; the total acreage offered was over 24,000,000.[7] In the tier of states from Minnesota to Louisiana, a number of the projects, like the Burlington and Rock Island roads in Iowa, were thought of as possible first stages of a transcontinental route. The 750,000 acres granted to the state of Michigan in 1852 for the Sault Canal resulted, as has been shown, in the construction of an improvement of importance to the entire Great Lakes region. Yet the number of these grants and the ease with which they were secured—until the movement was checked by the financial crisis of 1857—seemed to indicate that those who wished federal subsidy no longer needed to demonstrate that their projects possessed a uniquely national character.

By this time, however, the center of interest had been taken by a much greater project that was advocated with particular urgency on grounds of patriotism and national interest—the Pacific railroad. "Improvement of communication with the Pacific coast," as its historian has pointed out, had been "an issue in American politics" from the early days of the republic.[8] The first notable illustration was the Lewis and Clark expedition sent out by President Jefferson in 1803 to cross the Rockies from the Missouri River to the Oregon country. Jefferson's interests included the possibilities of both economic and territorial expansion as well as the increase of scientific knowledge, and the explorers were instructed to study trade routes not only for the Oregon fur trade but also "for the commerce of Europe, Asia, and America." Like the Gallatin Report which was presented five years later, the mission of Lewis and Clark illustrated the determination to use a knowledge of geography in the service of economic development.

In the decades that followed, western exploration was one of the functions of the Army, and a number of notable expeditions were made. The Army also did a substantial amount of road building in the West. The great early routes of travel to the west were, to be sure, for the most part discovered by private traders or trappers or by the "mountain men" who acted as guides to the emigrants' wagon trains. But Westerners had no difficulty in appreciating the economic fact that a road built by the Army for the transport of its own materiel would also serve for the carriage of settlers and their produce, and the political fact that it was easier to persuade the Washington authorities to build a road for "military" than for commercial purposes.[9] There were, however, some proposals for road making on a bolder and more frankly developmental scale; and Thomas H. Benton and others urged improvement of the Missouri and Columbia Rivers and the building of a turnpike between the two.

If this project resembled Gallatin's proposals for his pairs of Appalachian rivers, other rival schemes recalled the "necks of land" which he wished to cut. The geography of Central America and southern Mexico offered isthmian alternatives to the long routes overland. In the year the Erie Canal was opened, the Ambassador of the Republic of Central America invited the United States to enter into negotiations over the building of a possible canal by way of Lake Nicaragua. Some twenty-five years later, a company controlled by Cornelius Vanderbilt held the concession for construction by this route; and, though the canal was not built, transit was provided during the eighteen-fifties by a mixed land-and-water service. Another American company provided for a briefer period similar transit by way of the Mexican Isthmus of Tehuantepec, across which it hoped to build a railroad. More was accomplished on the Isthmus of Panama, which had provided one of the main routes of passage for the participants in the California Gold Rush; and the Panama Railroad ran its first through train from the Atlantic to the Pacific in 1855, the same year in which the Pennsylvania Railroad reached Pittsburgh. The United States government gave diplomatic support to these ventures and occasionally the services of an army engineer, but they received no financial subsidy.[10] This, indeed, seemed one of their great advantages to those who opposed the heavy governmental expenditures that would be required on the overland routes; and the shipping interests interested in the isthmian projects with their backers in New Orleans and New York and other ports, continued to play a part in influencing federal policy.

It was the transcontinental railroad, however, which became the main object of interest and political controversy. In 1845 petitions and memorials began pouring into Congress, and from then on there was hardly a time when one or more Pacific railroad bills were not under consideration. The settlement of the Oregon boundary in 1846, the acquisition of California and the Southwest in 1848, the discovery of gold in 1849, and the continuing growth of settlement in the Far West strengthened the political and economic arguments for national action.

Of the many schemes put forward, within Congress or outside, a number proposed the method of purely public enterprise. To some Westerners it seemed as natural for the government to build a railroad—at least "a rough, rapidly extended" one—as for it to build a post road or military road.[11] Senator Benton urged "a national highway; no stock-jobbing"; under his plan the federal government would have built a railroad and a common road running side by side from the Mississippi to the Pacific.[12] A proposal for the construction of a government railroad, using the services of private contractors, was before Congress in the 1857–58 session. Much later, at the close of the Civil War, the *American Railroad Journal* gave considerable space to a proposal to build the work under military auspices. Instead of disbanding the Army's sappers and miners, the government would keep them at work under their own trained officers. As the railroad was completed, the men would be given land along the way, like Roman legionnaires in a border province, and contribute to the settlement of the area.[13] The editor of the *Journal*, Henry Varnum Poor, at one time suggested still another variant, namely, construction by a board of directors appointed by the states. His belief was that the national government must provide the resources but for the sake of efficiency "should have as little connection as possible with the construction and management of the road." [14] None of these various measures for public enterprise came close to securing majority support, and the prevailing opinion was that stated by President Buchanan: "It would be inexpedient for this Government to exercise the power of constructing the Pacific Railroad by its own immediate agents." [15]

On the other hand, there were few serious proposals for construction of the road as an unaided private enterprise, though at one point in the discussion Benton reversed himself and declared that responsible capitalists would be prepared to build the railroad without subsidy. The great majority of plans were based on one form or another of partnership between the national government and private enterprise. Considering the vast stretch of unsettled country to be crossed and the physical obstacles of mountain and desert, it was the general opinion that no road could be built "without substantial aid, either in money or lands, from General Government." [16] And to most advocates, as to the Connecticut legislature in 1847, it appeared that the public lands were

> a fund peculiarly appropriate for defraying the expense of such an undertaking, as a small portion of them would furnish the necessary means, and the value of the remainder would be greatly enhanced thereby.[17]

The most discussed of the early projects, that of Asa Whitney, claimed the merit of being "purely an individual enterprise," which asked not a single dollar from the public treasury and would therefore "disembarrass" the undertaking "from sectional and constitutional objection."[18] Yet the Whitney plan rested, even more than most of the others, on the use of a great block of public land. What he asked was the assignment by Congress of a strip sixty miles wide from Lake Michigan to the Pacific Ocean. This he would buy at the price of sixteen cents per acre, or, in a later version, ten cents. The difference between this sum and the proceeds of the sale of the land would furnish the funds for construction and the profit, if any remained, for Whitney himself. The one element of subsidy lay in the sale of the land at the low price, and the difference in value would largely be created by the act of construction. Thus, said an advocate, "the whole capital required on the Whitney plan, lies at this moment sleeping in the land through which the road is to pass."[19] Once built, the railroad, unburdened by interest costs, would be run at rates controlled by Congress and sufficient only to cover the costs of operation.

In arguing for the proposal, Whitney and his supporters laid particular emphasis on the capture of the trade with China and the Indies and on the enormous profits to be made in the exchange of products with the ingenious and industrious millions of Asia. A House Committee declared that adoption of the plan "would make the United States the centre and axle of the commerce of the world," with the result—in a quick change of metaphor—that Europe would have to "bow to Asia, and Asia to Europe, across our bosom."[20] Whitney's campaign had an extraordinary effect in arousing interest in the Pacific project, and between 1845 and 1850 his plan was endorsed by eighteen state legislatures and many chambers of commerce and other civic bodies. Yet its chances of acceptance diminished as opponents on one side questioned the practicality of constructing a railroad by the means proposed and opponents on the other side insisted that the enterprise, if it should be successful, would put intolerable power in the hands of a single individual. A rival measure emphasizing private leadership was put forward by a group of Boston capitalists who were perhaps more realistic about costs and certainly frank about the need for subsidy. Their proposal was that the road be built by a company with resources of $100,000,000, of which all but $2,000,-000 was to represent a government loan. This received much less support than Whitney's plan, and still less attention was paid to a suggestion that the government might, as had been done in France and India, support a private company by guaranteeing a five percent return on its investment.

As the debate continued, discussion centered mainly on proposals for construction by a private company or companies with substantial government aid. With respect to subsidy there were two principal issues: First, how much public land was to be given? Second, should a financial subsidy be added to the land grant, and if so, how much? A series of measures considered between 1853 and 1859 gave somewhat varying answers. Land grants were proposed of forty sections per mile, twelve sections, and twenty sections, all on the alternate sections basis. In the first of these cases, there was to be no financial subsidy, except in Texas where the federal government held no lands. An amendment was proposed to substitute for the land grant a loan of $30,000,000. This was not accepted, but the later bills added loans to land. After earlier proposals of $20,000,000 and $15,000,000, a carefully drawn Senate bill of 1859 provided for a land grant of twenty sections per mile and a loan of $25,000,000 to be regarded as prepayment for the carriage of government freight and mail. Each of these measures, for all their emphasis on private construction, provided that, after a period of years, the road should revert to government ownership, in most cases without additional compensation. The recapture provision, however, was omitted from the Curtis Bill of 1860, which also proposed to offer to investors a larger financial inducement than any of the earlier measures. As originally put forward, the measure offered a land grant of only one section per mile; but the prepayment for transportation was to be $60,-000,000, and even this figure was raised by amendments providing for additional lines.

Agreement on terms of very substantial subsidy offered no great difficulty for the advocates of a transcontinental railroad. The sticking point was the selection of the route. Just as Atlantic seaports had projected rival ventures for the passage of the Appalachian Barrier, a similar array of Middle Western cities—on the Lakes, on the Missouri, and along the whole length of the Mississippi—put forward their claims to provide the eastern terminus of the Pacific railroad. Whitney's original plan was for a route from Milwaukee to Puget Sound, but there were many rivals. In a number of states, as the preceding Chapter has shown, east-west railroads were being pushed forward in the hope that they might serve as the "first links" of the transcontinental. The question, according to a Wisconsin paper, was: "Shall the upper West or shall the lower West be the great avenue of trade and commerce?" [21] As the Civil War approached, the bitterest controversy was between those who wished a road to serve the North and those who wished a road to serve the South.

Midway in the long debate, an attempt was made to place the selection of a route on a more objective basis. The Thirty Second Congress, as one of its last acts in March, 1853, asked the Secretary of War to conduct surveys to determine the most practicable and economical route for a railroad from the Mississippi to the Pacific. Opponents objected that this was a return to the dangerous precedent of the General Survey Act of 1824,[22] and it did indeed appear to represent the revival of an attempt at national planning on a broad geographic basis. Under this measure parties of topographical engineers were sent out to make reconnaissances and examinations of the five major routes around which discussion had principally centered—those of the 47th and 48th parallels by way of the Missouri and the Columbia; the 41st and 42d parallels following the path of the wagon trains up the Platte Valley and through South Pass; the 38th and 39th parallels favored by Senator Benton and his explorer son-in-law, John C. Fremont; the 35th parallel by way of the Santa Fe trail; and the

32d parallel by way of El Paso, Texas, and the Gila River. Each of the five exploring parties presented estimates of length, grades, altitudes, costs, and—to some degree—opportunities for settlement along its route. These findings were compared and critically examined by the professional officers in the Corps of Topographical Engineers, who modified the original estimates. The whole was presented to Congress in 1855 by the Secretary of War, with his own statement of the conclusions.

The 38th-parallel route, on which no pass was found at less than 10,000 feet, was pronounced too costly to be considered. Each of the other four routes was reported as practicable. These broad judgments have been confirmed by later developments. The Northern Pacific, the Union Pacific, the Santa Fe, and the Southern Pacific all follow in general and often in close detail the four recommended routes, except that the Union Pacific succeeded in finding a better alternative to South Pass. On the other hand, though the Denver and Rio Grande crosses the Rockies at about the 38th parallel, no railroad has ever continued the line across the Sierras at this latitude.

More debatable, and less fully accepted, was the Report's rating of the merits of the practicable routes. The northernmost had the advantage of easy grades and the lowest "reported sum of the ascents and descents." The central route aimed directly for San Francisco, the most important goal. The Sante Fe route made use of the "interlocking tributaries of the Mississippi, the Rio Grande, and the Colorado of the West" and appeared to go "through or near more cultivable areas" than the others. Comparison of the four, however, was said to show "conclusively . . . that the route of the 32d parallel is, of those surveyed, the most practicable and economical." It was the shortest line to the Coast; its highest pass was under 6,000 feet; and much of the route lay over an almost level tableland. As a result, its estimated cost was "less by a third than that of any of the other lines." [23]

Since the Secretary who presented the Report was Jefferson

Davis of Mississippi, this unqualified finding in favor of the southernmost route has often been attributed to sectional prejudice. Yet almost all of the officers who led the surveys or reviewed their findings were Northerners, and the charges of major distortion have not been proved. When the reviewing officers added twenty percent to the original estimates of cost by way of the Missouri and Columbia Rivers, it seems probable that they were correcting rather than introducing a bias. On this route the explorations had been conducted by the enthusiastic Governor of the Washington Territory, I. I. Stevens, who had proclaimed his conviction that Nature had "indicated the northern pathway." [24] When railroads were finally built, under changed technical conditions, it was found that the actual distance was a little greater by the Gila River route and a little less by the other routes than the Report had estimated. The Secretary's final conclusions, moreover, emphasized the southern route's advantage in distances to "Atlantic and Gulf seaports" without mentioning the less favorable comparison of distance to Chicago or St. Louis.[25] Yet these were hardly conclusive points, and the Surveys taken as a whole provided a notable addition to the knowledge on which Congress might base a rational choice.

What is less clear is that they had any substantial effect on congressional decisions, except to discourage proposals to build westward from the state of Missouri. While the exploring parties were in the field, the administration was already negotiating with Mexico to buy enough land to bring the Gila River route entirely within the borders of the United States, and the Gadsden Purchase treaty was ratified the year before the report on the Surveys was presented to Congress.[26] If this indicated the strength of prior commitment to the southernmost route, the partisans of the northern and central routes were by no means willing to abandon their own schemes and accept Secretary Davis's recommendation. Remarking that this "true son of the South" had shown "a little Southern proclivity in his report," [27] they proceeded to draw from the surveys themselves arguments for the practicability and advantages of their own proposals.

These divisions were reflected throughout the course of the debates. The earliest major bills proposed two railroads, one to the northwest following Whitney's plan and the other by a southern route. Senator Gwin of California suggested a single trunk line by way of Texas with a number of eastern branches. In 1854–55 Senator Douglas sponsored a bill providing for as many as three railroads, if proper bids were received, to start from the western boundaries of Wisconsin, Iowa or Missouri, and Texas respectively. An amendment reduced this to one trunk line by a central route with branches to Memphis and Lake Superior. In later Congresses, the characteristic pattern was the introduction of a bill for construction by the central route, an immediate countermove to add a southern road, and often an attempt by a smaller but highly persistent group to add a road by the extreme northern route. Thus in 1859–60, the Curtis Bill, which originally provided $60,000,000 for the central route but to which successive amendments added $36,000,000 for a southern road and $25,000,000 for a northern road, passed the Senate—but not the House—after six Southern states had already seceded. At no time was the Congress prepared to leave the choice of route either to the administration or to what its members called the "eagle" or "lynx eye of capital." Though a large majority of senators and representatives were for years in favor of some form of federal aid to a Pacific railway, the conflicts of local and sectional interest continued to prevent agreement on any specific national plan.[28]

Secession removed from Congress almost all the partisans of the southern routes, as well as the most consistent opponents of federal aid, and supporters of the central route now held a clear majority over those of the northern. Moreover, the old economic and political claims for the railroad were reenforced by new arguments for binding California more closely to the union and for protecting the Pacific Coast against dangers such as those suggested by Maximilian's conquest of Mexico. Against these the argument that building a transcontinental railroad would represent a distraction from the war effort appeared to carry little weight. Yet the "wrangle of local interests" continued to make decision difficult, and the exasperated Delegate from New Mexico complained that the professed advocates of a Pacific railroad would not support any particular bill "unless it starts in the corner of every man's farm and runs through all his neighbors' plantations." [29] The final effect of the Pacific Railway Acts of 1862 and 1864 was to fix the eastern terminus of the main line at Omaha, Nebraska. This decision represented the victory of the "Iowa-Chicago interest" supported by New York over the "Kansas–St. Louis interest" supported by Cincinnati, Baltimore, and Philadelphia.[30] But agreement on a single line could not be reached, and the legislation as adopted made the Omaha route only one of five eastern branches entitled to federal aid. The result, said one critic, was to turn an artery into a sprinkler.[31]

The provisions of the two decisive measures are well known. The 1862 act provided for the organization of the Union Pacific Railroad Company to build the main line as far as the California border and one or more of the eastern branches, and it authorized the Central Pacific Company of California to build from Sacramento to the state line. In case one of the companies reached the meeting place before the other, it was entitled to continue building until the junction was made. The subsidy provisions followed in general the precedents worked out in the bills of the preceding decade. For the main line and the authorized branches, aid was to consist of ten sections per mile and a first mortgage loan of $16,000 per mile for the eastern and western parts, of $48,000 per mile for 150-mile stretches in the Rocky and Sierra Nevada Mountains respectively, and of $32,000 per mile in the Great Basin between them. Some of these subsidy bonds were to be held back until the entire road was completed, and their total for the main line was not to exceed $50,000,000. All payment for services rendered the government and at least five percent of net earnings after completion were to be applied to repayment and interest on the loan. Before subsidy could be extended for any section of the road, its satisfactory completion must be certified by government commissioners; and the government was to appoint two of the fifteen directors of the Union Pacific.

Under this act some construction was carried on by the Central Pacific and Leavenworth companies. The Union Pacific obtained subscriptions of a little more than $2,000,000, the sum required for the organization of the company, and the ten percent paid in cash was supplemented by short-term borrowing. Of the sums thus raised, a part was spent in starting construction from Omaha and a larger part in the attempt to obtain more favorable terms in Washington.

By 1864 it was clear that the conditions of the original grant were not sufficiently attractive to induce capitalists to build a railroad across the plains and mountains. Most members of Congress remained anxious to secure its construction, but they differed as to the appropriate means. A minority drew the conclusion that the government should build the road on its own account. Congressman Pruyn of New York estimated that capitalists would "contribute two percent only of the money, and no part, of course, of the real estate. In other words," he concluded, "the Government in fact built the road, and ought to control and own it." But the majority held to what one congressman described as the established principle "that the Government ought not to appoint their own Federal agents for any great work of public improvement." The alternative was to improve the terms offered to private enterprise. What Congress proposed, it was said, was "to do enough, and only enough, to induce capitalists to build the Pacific railway." [32]

The 1864 act was framed in the belief that these inducements should indeed be substantial. The land grant was doubled to twenty sections per mile. The provision for reservation of bonds was repealed, and the installments of subsidy were to be issued on conditions that would reduce the amount of working capital required. Half of the government's payment for services was to be made in cash and only half set off against the loan. Most important of all, the government's claim was reduced to a second mortgage, and the companies were permitted to sell their own first mortgage bonds to private investors in the same amounts and at the same times as the government bonds. Stock of the Union Pacific Company was to consist of $100,000,000 in shares of $100 apiece, which must be paid for in cash, and the books were to be kept open until the entire amount was subscribed.

These 1864 proposals seemed to one congressman "the most monstrous and flagrant attempt to overreach the Government and the people" that had ever been made, and some though not all the participants in the debate were convinced that reducing the government's claim to a second mortgage meant that it would never be repaid. Under these new conditions, it appeared urgent to some legislators to provide additional assurance that the government's purposes would really be carried out. One representative attempted, without success, to provide that construction contracts made by the companies should require approval by the Secretary of the Treasury as a test of their good faith. Others regarded it as particularly important that the "gentlemen . . . representing the interests of the Government in this corporation should have very large powers." [33] In response to this feeling, the 1864 act increased the number of government directors in the Union Pacific to five, in a board of twenty, and provided that one of these should be a member of each of its standing or special committees.

The sequel is familiar, at least in its general outlines, to all students of American history. The nearly eighteen hundred miles of railroad connecting Omaha and Sacramento were completed in 1869. The advance of the Central Pacific's Chinese laborers from the west and the Union Pacific's Irish laborers from the east and the driving of the final gold spike at Promontory Point have become a permanent part of American folk tradition. The physical achievement, in the face of Indian attacks, difficulties of supply, and all the obstacles of mountain and desert, was a remarkable one, and it included the notable discovery by the Union Pacific's engineer of a pass which provided an easier transition from the Plains and easier grades over the Rockies than the South Pass of the emigrant trains. The advances made by the government on the main line were eventually repaid, though only after years of litigation and controversy. As a developmental measure, the success of the undertaking has never been seriously questioned. Though its impact on the economy as a whole was less than that of the Erie Canal and the trunk-line railroads, it has often been regarded as the greatest, as it was certainly the most costly, of all the publicly supported improvements of the ninteenth century.

The story of the Union Pacific's financing is also familiar. The government bonds sold at par or close to it, and the first mortgage bonds were disposed of at small discounts. On the other hand, the amount of "venture" capital that came forward was very small indeed. The law's requirement that subscriptions should be in cash only was evaded by the issuance of stock as part of the payment on construction contracts. The same group of persons held the stock of the Union Pacific and the Credit Mobilier of America, a corporation organized to receive and distribute the profits of construction. Even before the road was completed, they managed to return to themselves through the Credit Mobilier almost all the cash they had put in; and within a brief period they paid themselves dividends in Union Pacific securities, which depended for their value on the prospects of the road but in current prices were worth three or four times their original investment. At the time when these profits were being made, shares of Credit Mobilier were distributed to a number of members of Congress, who made little or no cash payment on them and whose obligations were soon canceled by dividends greatly exceeding the par value of the stock. The same pattern of small initial investment, large profits through construction contracts to insiders, and the distribution of favors in Washington was repeated with variations in the case of the Central Pacific.

Accounts of the episode have differed greatly, not so much by denial of any of these statements as in the assessment of the relative importance of the various points. From the point of view of national policy concerning internal improvements, the most significant questions may be stated as follows: Assuming that the considerations of economic development and political unification were sufficiently cogent to justify government expenditure for building a Pacific railroad before it would be undertaken by unaided private enterprise, were such lavish terms of subsidy needed to secure private construction? In other words, did government policy meet the specification of doing "enough, and only enough, to induce capitalists to build the Pacific railway"? And if so, were these terms too high a price for the government to pay to avoid having to do the work by its own "direct agency"?

Analysis of the inducements to capital, and of their effects, must distinguish between loan capital and equity investment. With respect to the former, there was no serious difficulty. Private investors paid $23,718,000 for $27,213,000 of the six percent first mortgage bonds of the Union Pacific. This part of the financing, since it involved no scandal, has been little discussed; but it is evidence that reduction of the government's claim to a second mortgage had its intended result and that the investing community came to believe that the enterprise would not be a complete failure. Investors also paid more than $27,000,000 for the government's six percent bonds. Smaller sums were raised by seven percent bonds on the security of the land grant and by ten percent income bonds, both of which sold at heavy discounts. The amount raised by these various issues was about $63,500,000.

Since the total cost of building the road was about $59,000,000, including some $2,500,000 of discount and interest on short-term loans,[34] the contribution of equity capital was small indeed. Those who controlled the enterprise were able to do so on the basis of an investment of about $4,000,000 in the stock of the Credit Mobilier. After the process of construction was started, this money served as a revolving fund to tide over the period of perhaps four months from the time work began on a section of the road until the bonds for that section could be issued. Most of this sum, in addition to a large amount of the securities of the Union Pacific, was taken out within a year after the completion of the road. The total payments from the railroad company to the Credit Mobilier, taking the securities at face value, were well over $90,000,000. Since the construction company paid less than $51,000,000 to the subcontractors for building the road, its nominal profits were more than $40,000,000. Most of this, however, was taken in common stock, which was currently worth perhaps thirty cents on the dollar though it was later to rise to par. In terms of current cash value, the profit has been conservatively estimated as $14,000,000 or $15,000,000.[35]

Even this modest and well-rewarded amount of "venture" capital, however, was difficult to raise. Prominent railroad men unconnected with the Credit Mobilier testified that they had regarded the Pacific railroad project as "one of the maddest material ventures" and "beyond the range of ordinary prudence . . . except as a government measure." [36] On the other hand, an investigating committee of the House of Representatives reached the conclusion that the promoters "differed from other capitalists not in taking a risk, but in having discovered that the road could be built at vast profit without risk." [37] Yet investors were slow to make or to credit the discovery. Thomas C. Durant had the Credit Mobilier device ready at the outset, having bought a convenient Pennsylvania charter several months before the 1864 act was passed; but he and his friends could not raise the first $2,500,000 for its stock without calling in Congressman Oakes Ames and a group of Massachusetts investors. By early 1867, the Evans Pass route had been discovered and adopted, more than three hundred miles of road had been built, and profits on the first construction contract were being distributed. Yet the final subscription of $1,250,000 could be raised, as Ames and his associates testified, "only with the aid of an enormous bonus" in Union Pacific first mortgage bonds.[38] On this point their behavior confirmed their testimony. If capital had been easy to obtain, they would hardly have sold land-grant bonds in 1866 at a rate to yield over eleven percent or continued until the middle of 1867 to borrow short-term funds at rates ranging from one to one and a half percent per month. It was not until the summer of 1867 that they felt sufficient confidence in the market to begin sale of the first mortgage bonds. But by this time Durant had fallen out with the Massachusetts group and was using the injunction weapon in an effort to block the action of the Credit Mobilier. Even when it became clear that large profits could be made, there remained uncertainty as to which group would obtain them.

Certainly there were no further doubts about profitability once the first mortgage began to sell and the two factions had agreed on the final contract.[39] Otherwise Oakes Ames would not have used Credit Mobilier stock so freely to win friends in Congress, at a time when no Pacific railroad legislation was under immediate question; and the financier Jim Fisk would not have attempted to obtain an interest in the company by bringing suit on slender legal grounds. Apparently, also, the prospect of large profits was recognized by suppliers and contractors in the later stages of construction; and the president of the Credit Mobilier declared: "We were plundered all along the line by everybody, who seemed to think that we were fair game." [40]

Construction costs were also increased by the provision in the legislation which allowed the Union Pacific and Central Pacific each to build on until it met the other.[41] Both companies were so eager to obtain the subsidy for additional mileage, and also to win the trade of the Mormon settlements, that they sacrificed economy to speed, working in the high mountains through the worst of the winter and sometimes even laying temporary track on the snow. There were, moreover, two hundred miles on which grading was done by both companies, on parallel lines, before the junction point was finally determined.

In most respects the story of the Central Pacific parallels that of the Union Pacific. The physical difficulties and the physical achievement were comparable. The business community was no less skeptical of the practicability of the enterprise, and venture capital was perhaps even more difficult to obtain. On the other hand, public aid was given on an even more generous scale. Together with the Western Pacific, which continued the line westward from Sacramento to San Jose, the Central Pacific received a total of nearly $28,000,000 in subsidy bonds and issued the same amount in first mortgage bonds. These totals were slightly higher than for the Union Pacific; though the mileage was less, more of it was in mountain and Basin territory. In addition, the two western roads received both state and local aid. The state of California, in spite of a constitutional provision against aid to corporations, agreed to pay interest for twenty years on $1,500,000 of seven percent Central Pacific bonds. The companies received also as subscriptions or donations $1,600,000 in city and county bonds. They appear to have sold the federal bonds at approximately the same discount as the first mortgage bonds, but received about $41,500,000 for the two; and they obtained an additional $2,000,000 by disposing of the local and state-aid bonds at about two thirds of their face value.[42]

In the construction of the Central Pacific, Charles Crocker and Company, and later the Contract and Finance Company, played a role corresponding to that of the Credit Mobilier. In this case ownership was even more closely held, and the entire control of the railroad and the construction companies was in the hands of four men—Leland Stanford, C. P. Huntington, Mark Hopkins, and Crocker himself. For this reason, and because the books of the Contract and Finance Company were reported lost at the time of congressional investigation, it is even harder than in the case of the Credit Mobilier to compare investment and return. Since almost no cash was paid in on the Central Pacific stock, and since the associates supported the contracting companies by personal loans rather than by subscribing to its stock, the total amount of bona fide equity capital was considerably smaller even than that of the Credit Mobilier. On the other hand, allowance must be made for the money loaned by the associates and, at certain difficult times, borrowed on their personal credit. The government and the first mortgage bonds seem to have substantially covered the cost of construction. When this was completed, the associates appear to have been left with about $38,500,000 in Central Pacific stock, and possibly certain other assets, as clear profit. The stock was of course a speculative asset, depending for its value on the future of the road; but a careful student, suggesting a valuation of fifteen or twenty cents on the dollar, concludes that it represented, "on the most conservative calculation, 500 or 600 percent for an investment which probably did not exceed $1,000,000, over a period of six years." [43]

For all the railroads affected by the laws of 1862 and 1864, the government appointed commissioners whose duty it was to certify satisfactory completion of the sections of road on which subsidy was claimed. In the case of the Union Pacific, but not of the others, there were government representatives on the board of directors. It may therefore be asked how effective were these "gentlemen representing the interests of the government." Among the commissioners, one individual demanded and received from the Union Pacific a bribe of $25,000 for certifying a particular section. In a number of cases, commissioners insisted on changes in construction; and officers of the Credit Mobilier complained that some of these modifications, made to satisfy men inexperienced in railroading, were unnecessary or even harmful.[44] In 1869, after criticism of the companies had begun to mount, a "commission of eminent citizens" was appointed by the government to examine the completed roads. These men suggested certain improvements, which were carried out during the subsequent five years; but their general conclusion was that the two roads provided "a good and reliable means of communication between Omaha and Sacramento." With respect to the Union Pacific, they declared that its location showed "a thorough knowledge of the features of the country." [45]

The role of the government directors continued until the subsidy bonds were paid off in the reorganization of 1897. Some were men of competence and distinction, such as Charles Francis Adams; others were clearly political appointees with no particular qualifications for the position. Their annual reports to the Secretary of the Interior [46] show that they exercised independent judgment on a number of occasions and played a somewhat useful part in the negotiations over final settlement of the government's claims. Of greater interest is their performance during the controversial period of construction. One of the early government directors obtained a block of Union Pacific and Credit Mobilier securities, in the year of most lavish distribution, at much less than market price and in violation of the terms of his appointment.[47] On the other hand, another government director, himself an engineer, took an active part in the work of the field parties and in the decision to recommend the Evans Pass route.[48] The report of the government directors presented in June, 1956, contained some criticism of the original Credit Mobilier contract; and four of the five government directors were recorded as voting against one of the later construction contracts. One of them also claimed credit for initiating a successful fight against a fraudulent arrangement for the purchase of coal from a company controlled by some of the promoters. Sometimes the government directors complained of their difficulty in obtaining information about the affairs of the railroad.[49] In any case it is clear that they made no objection in principle to the device of the construction company and that they had too little power in the board to prevent the lavish distribution of profits. Nor is it at all certain that they would have been supported by the Administration if they had protested on either point.

Testimony given before the investigating committees contained criticism of the government directors from two points of view. Congressmen pressed the charge that they had shrunk from the exercise of the more unpleasant parts of their duty, and drew from one of them the admission that he had walked out of a special committee investigating certain "legal expenses" because he "did not want to know anything about it." "The Government directors," said a company witness, "are the most squeamish men that ever lived about taking any responsibility." On the other hand, spokesmen for the railroad regarded it as an imposition that the Union Pacific, unlike the other companies, should have to submit to the presence on its board of men who were "not likely to know anything about the construction of railroads" and who had no interest in the profits of the company. They said that this requirement was one reason for the reluctance of capitalists to invest in the enterprise. Even after the construction period was well over, the president of the Union Pacific described the arrangement as "a most unfortunate circumstance for this company." [50] A mixed board of public and private directors, so common an arrangement in the early days of railroad building, was now regarded as an anomaly; and even critics of the company argued that settlement of its financial obligations would have as one advantage "a speedy and absolute divorce" permitting the government to "cut loose from all participation in railroad management." "The sovereign," said the Pacific Railway Commission of 1887, "should not be mated with the subject." [51]

The 1864 act was the last to offer financial subsidy to a railroad, and the board of the Union Pacific was the last to which the federal government named directors. On the other hand, grants of land were made for each of the transcontinental roads which followed the three other routes found practicable by the Pacific Railway Surveys. The advocates of the northernmost route secured aid for the Northern Pacific in a bill signed on the same day in 1864 as that of the Union Pacific. For the route of the 35th parallel, an act of 1866 provided aid for the Atlantic and Pacific, with the western end or "branch" to be built by the Southern Pacific of California, a company controlled by the same interests as the Central Pacific. In 1871 the last of these grants was made for the Texas and Pacific to build on the route of the 32d parallel. Under these three measures land was given even more generously than before, whereas direct financial aid was refused. The new provisions called for forty sections per mile in the territories (double that of the Union Pacific grant) and twenty sections within the states.[52]

Neither these subsidies, nor the Union Pacific's demonstration that a transcontinental railroad could be built and operated,[53] resulted in rapid completion of the competing roads. As in the case of the state improvement programs of 1836, part of the difficulty lay in the fact that the enterprises were initiated just before the onset of a major depression. The potential resource of its huge land grant did not prevent the bankruptcy of the Northern Pacific in 1875, and the road was not completed to the coast until 1883, only a year before the Union Pacific reached the Northwest by means of the Oregon Short Line. The Atlantic and Pacific also failed, leaving its grandiose name for later appropriation in another field, after building from the end of one of Missouri's state-aided "first links" for a short distance into Indian Territory. Its rights to further land grants were eventually acquired by the Atchison, Topeka, and Santa Fe Railroad, and in 1883 the Santa Fe reached The Needles on the Colorado River, meeting the Southern Pacific there and completing a through connection by way of the 35th parallel.

The story of the southernmost route is still more complex. Again the Southern Pacific preempted the western end, having built a line to Yuma, the only other practicable crossing of the Colorado, even earlier than to The Needles. But at the other end the Texas and Pacific, in spite of federal and state aid, was making very slow progress across the Texas plains. In this situation the Southern Pacific, though not entitled to land grants east of the California border, decided to build on across Arizona and New Mexico over the favorable tableland described by the engineers of the Pacific Railway Surveys. The junction of the two roads was finally accomplished east of El Paso, Texas, completing a route by the 32d parallel; and shortly afterwards the Southern Pacific built and acquired an independent line of its own to New Orleans.[54]

Construction of these portions of the Southern Pacific without federal aid may well be taken as an indication that the difficulties of crossing the continent could no longer provide a cogent argument for national subsidy. Yet during this period, and for each of these three routes, attempts were made to add financial subsidy to the aid in land. The method proposed was that of a government guarantee of interest on the bonds of the companies. A House committee in 1869, noting that Louis Napoleon had used this form of subsidy successfully in France, urged its application on a second and a third transcontinental railroad. Among the many arguments for aiding in their construction, the Committee declared that "the North Pacific road will acquire for us the British possessions" and that "the Southern Pacific road will annex Northern Mexico." [55]

A much more serious effort was made from 1874 on to secure a government guarantee for the bonds of the struggling Texas and Pacific. Under the vigorous leadership of its president, Thomas A. Scott, a popular campaign was conducted that rivaled Asa Whitney's, securing endorsements from almost every southern legislature and winning considerable support even in the North. A recent study has demonstrated that commitments by northern political and financial interests to support the guarantee played a part in the complicated negotiations as a result of which the Democrats accepted the seating of Rutherford B. Hayes in the disputed presidential election of 1876 and the Republicans agreed to the ending of Southern Reconstruction. Even this support, however, was not enough to carry the measure, partly because Scott's Texas and Pacific lobby was opposed behind the scenes in Congress by Huntington and the Southern Pacific [56] and partly because of a general feeling that the time had passed for obtaining financial aid from the federal government.

The time had passed, also, for what had seemed the less controversial practice of giving subsidies in land. The period in which grants were made directly to railroad corporations continued for ten years after its inauguration in 1862, and the total acreage authorized for distribution was more than 116,000,000. Of this 24,000,000 acres were for the first transcontinental road and its branches, 44,000,000 for the Northern Pacific, nearly 23,000,000 for the Atlantic and Pacific, 13,000,000 for the Texas and Pacific, and about 7,750,000 for the two western connections provided by the Southern Pacific. The rest of the total represented five other grants in the West and Far West. Meanwhile, the older practice of grants to the states for railroad aid continued, though on a smaller scale. The list of measures from 1862 to 1872 is a long one, but many merely reaffirmed or reassigned grants previously made for railroads that had failed to meet the required conditions. The total of new authorizations was about 11,000,000 acres. Of this nearly 4,000,000 were granted to the new state of Kansas, mainly for the benefit of the Atchison, Topeka and Santa Fe. Congressmen voting for the Texas and Pacific measure in 1871 announced that they shared the general feeling of the country against further land grants to railroad corporations but justified their action on grounds of belated justice to the South. In the following year the appropriation of federal lands for railroad construction terminated with a final state grant, to Michigan for the use of the Chicago and Northwestern.[57]

These actions brought to an end the era of national subsidy. The grants already made, however, continued for some decades to require further legislative and legal action. The subsidy bonds gave rise to a series of controversies and kept the government involved in railroad "management" until the time of repayment. The courts early held that interest on the bonds, except in so far as it was covered by half the government's charges for transportation and the five percent of net earnings, need not be paid until the principal fell due at the end of the thirty-year period. Repayment of these large sums appeared highly doubtful, as the companies paid out earnings freely in dividends; legislation was adopted in 1878 in the attempt to force them to provide a sinking fund sufficient to meet these obligations. The final outcome was more favorable to the government than many observers had expected. Reorganization of the Union Pacific system after receivership resulted in 1898 in the payment of principal and interest on the Union Pacific's own bonds and of principal but not interest—a difference of some $6,000,000—on the bonds of the Kansas Pacific, the successor to one of the eastern branches. In the following year, negotiations assured the payment, subject to brief delay, of both principal and interest on the bonds of the Central Pacific and the Western Pacific.[58]

The land grants also required continuing oversight and a number of administrative and legislative decisions. Passage of a railroad land grant did not automatically transfer title to the company nor even determine precisely the acreage involved. All grants were made subject to the prior sale or homesteading of the land falling within the boundaries established by the law, though most of them permitted the companies to select compensatory or "lieu" lands within a wider belt of territory. If this acreage was also preempted, the company received less land than its mileage had "earned." Without this restriction, the grants would have covered an area that has been estimated as between 180,000,000 and 215,000,000 acres rather than the figure of 155,000,000, which is customarily used. Of this the railroads in the end received 129,000,000 acres. The difference is explained by the fact that the companies lost the benefits of the grants if they failed to construct the mileage for which the grants were given or if they failed to do so within the time set, though these limits were often extended. Sometimes the provisions were evaded, and the courts held that forfeiture under them required specific congressional action; but such measures became increasingly frequent after the first case in 1870. Finally, an act of 1890 made forfeiture general for all grants that remained "unearned" by actual construction. The largest grant revoked was one of 13,000,000 acres which had been given to the Texas and Pacific Railroad, and the only federal land actually received by this system was some 1,000,000 acres granted to its eastern connection, the New Orleans Pacific. The Santa Fe received the substantial amount of 10,000,000 acres, in addition to 3,000,000 transferred by the state of Kansas, but this was less than the Atlantic and Pacific grant had appeared to cover. On the other hand, the roads included in the Union Pacific act of 1864 received substantially the amount estimated; and the Northern Pacific, in spite of its slow start, patented the largest amount of all, some 39,000,000 acres.[59]

The era of national subsidy was not an era of national planning. To say that a measure tended to create a "system" of internal improvements was still an effective argument for the opposition. In the Pacific Railway Surveys and in other explorations, the government performed the more modest function of providing data on which its own decisions and those of private promoters could be based. In the selection of particular projects for federal assistance, there were a variety of rationales and varying degrees of rationality. The first railroad land grant, that for the Illinois Central and the Mobile and Ohio, could be defended as aiding an intersectional route of national importance. In most of the cases that followed, no such claim could be made on behalf of the particular enterprise, and the choice of roads and routes depended more on the pressures of local and regional interest than on any consistent basis of judgment regarding their merits. They might, however, be defended on the general ground that the public domain could well be used to encourage railway building by private enterprise in the newer and less developed states.

For the transcontinental railroads, on the other hand, the case for federal subsidy was more specific and much more cogent. Since the routes lay mainly through the territories, the states could not take the responsibility. The great distances of sparse settlement or of none at all presented an economic difficulty which everyone recognized as formidable; the physical obstacles were greater than any previously encountered in the advance across the continent. If it is recalled that the most serious attempt at national planning, as well as some of the most successful cases of state and municipal promotion, took place when the Appalachian Mountains stood in the way of economic development, it should not be surprising that the most extensive use of national subsidy occurred when the physical barrier was that of the Rockies and the Sierras.

To overcome this barrier by the building of the first transcontinental railroad was an undertaking of major importance, and there can be no doubt that it accomplished its intended purpose in the promotion of economic development. "When the Government subsidized that road," said a Union Pacific president, "the Government did what any private party, under similar circumstances, would have afterwards considered the monumentally successful business operation of his life." [60] This was an *ex parte* claim; but it may well be conceded that the advance to the companies of some $60,000,000 in bonds and the large land grants were not too high a price to pay—if it was necessary—for the very great benefits obtained by the nation as a whole.

It is, however, by no means clear that this particular price had to be paid, or that the means employed were essential to the results obtained. They served to get the roads built by private management, but they did so at heavy social costs. Small groups of insiders venturing very little capital of their own were able to build the roads with the money of the bondholders and the government, and to pay themselves quick and extraordinary returns out of the profits of construction. In the case of the Union Pacific, they were able to take out large gains even before construction was completed. In that of the Central Pacific, a group of four men who had put up even less of their own money remained in sole control of a great railroad. As a result both companies were left with greatly inflated capitalizations, which reduced their ability to provide efficient service at reasonable rates. Nor can a judgment rest on economic factors alone without taking account of the effect on the national life of the notorious evasion of law in the financing of construction and the flagrant use of financial means to secure influence in Congress.

Equity capital certainly could not have been attracted for investment in the transcontinental railroads without the prospect of very large profits. Yet with what we know now of the costs of actual construction, it is clear that such prospects were in fact—if not in expectation—held out under the terms of the 1864 act, even if these terms had been strictly enforced, so long as they included the essential concession of reducing the government's claim to a second lien. If the companies had dealt directly but at arm's length with the actual working contractors, they would presumably have secured lower prices than those charged when the Credit Mobilier appeared fair game to all. Consequently, the costs of construction would have been less, and the roads would have begun their transportation operations with a smaller debt and more adequate working capital.[61] Under these circumstances it appears probable that the investors would have been rewarded at least as generously and as promptly as those who in an earlier period contributed venture capital to the Pennsylvania or the Baltimore and Ohio. But since the terms of the act, which a court later described as "a proposal to enterprising men," [62] offered a loophole for the quicker profits of the construction contracts, they were hardly likely to call forth investors who looked for their returns to the more prosaic method of railroad operation. It is, moreover, most doubtful that a substantial amount of such investment would have come forward in any case in the middle sixties, in view of the widely held belief that the undertaking was impracticable. But if such capital could not be obtained, and if the improvement was nevertheless regarded as a matter of national importance, it would appear that its construction as a public work, if the climate of opinion had permitted, would have been preferable to the terms on which private entrepreneurship was in fact secured. "In any other country," said Charles Francis Adams, an undertaking like the Union Pacific "would have been built by the Government as a military road." [63] The arguments for mixed enterprise, which had played so large a part in the earlier development of American improvements, could hardly be stretched to cover so glaring a disproportion between the contributions of government and of equity capital.

A different set of judgments applies to the policy of land grants as it was applied both to the transcontinental and to the other roads. From the government point of view, this method of subsidy had certain obvious advantages. The argument that the government as a prudent proprietor might well dispose of part of its holdings in order to enhance the value of the rest had the double merit of making economic sense, so long as the land was regarded as a fiscal resource, and of quieting constitutional doubts of federal authority to assist internal improvements. An official investigator for the British government, Captain Douglas Galton, described the method as an admirable invention and recommended its adoption in the British colonies. It seemed to him to have the particular merit of permitting the government to subsidize the railroad companies without imposing on them the penalty of interference with management.[64] On the other hand, the government's interest, which was to get the improvements made, could be protected by the conditions of the grants, provided that the measures were well drawn and effectively administered. The principal reason for the popularity of the method, however, was a much simpler one.

It was a form of subsidy which appeared to be costless. By its means, legislators could accede to the demands of promoters and localities without the unpleasant necessities of raising taxes or borrowing money or incurring current deficit. Future development, it appeared, could be promoted by using a resource which had little or no present value.

Unfortunately, however, the advantage of the method to the railroads was affected by the same consideration. How was a commodity which had no current value to the government to be transmuted into an asset which would make funds immediately available for railroad construction? The alchemy was, of course, to be performed by the building of the railroad itself; but the process was one in which the timing was far from perfect. The Illinois Central ran through an area in which a large amount of land of high agricultural quality had remained unsettled; and the company was able to sell almost half of its lands during construction and the year following for a total amount of over $15,000,000. On the other hand, land sales contributed only a little more than $100,000 to the nearly $50,000,000 raised for the construction of the Central Pacific. Of the two cases, that of the Illinois Central was the more exceptional. With slender immediate returns from sales of land, most of the land-grant railways attempted to anticipate their value by floating loans with the land as a security. Of the total nominal capitalization of the Union Pacific, for example, a little less than a tenth consisted of land bonds; but these paid seven percent interest as compared to the six percent of the first mortgage and subsidy bonds and were sold at much greater discounts. A prudent investor had to take into consideration not only the future value of land and the period within which it might be realized but also the uncertainties as to how much acreage the railroad would "earn" and how much it would finally receive. The amount that could be realized on the land grants at the time the funds were most needed was thus sharply limited. On the other hand, once the railroad was built and prospering, they would give the company additional returns at a time when there was no longer any justification for subsidy.[65]

In these respects, the effect of subsidy by land grant differed sharply from the effects of subsidy by loan. In a financially successful improvement of the developmental type, costs exceed receipts until the improvement is completed and has begun to have its effect in promoting development, but afterwards substantial revenues are obtained. The loan confers its benefit in the former period when it is needed and places most of its burden in the latter when it can be borne. The land grant, on the other hand, divides its benefits uncertainly between the two.

The railroad land grants have also been criticized and were often deeply resented on the ground that they deprived settlers of land which they might otherwise have received free or at a lower price. The Homestead Act was also "a proposal to enterprising men," though offering less princely rewards, and the same land could not be used for both purposes. Yet the conflict in interest was by no means a complete one. The railroad grants and free land were both expressions of the policy of using the public domain as a means for encouraging rapid settlement rather than as a fiscal resource for the government. Under the terms of the legislation, the grants given to the companies consisted of land which no settler and no speculator had yet found it worth while to acquire. On the western frontier, the presence of a railroad would in general do more to make settlement possible and profitable than a price increase of a few dollars an acre would do to make it unprofitable. Whether the increase would in fact be moderate depended, of course, on the policies adopted by the individual company. The land-disposal and colonization activities of a number of the land-grant railroads have recently been the subject of intensive study, and the choice facing them has been stated in the following terms: "As land companies they were concerned to sell their lands at the highest possible price and with the least possible expense. But as transportation companies they sought to have their areas developed as speedily as possible by settlers who would provide traffic." [66] The latter consideration would impel them to seek out colonists and to offer them reasonably attractive prices and conditions of sale. Some of the laws attempted to strengthen this tendency to providing that the land must be sold within a given period or else forfeited; others tried to prevent further speculative gains by providing that it could be disposed of only in quarter-section lots. Other aspects of the laws and their administration, however, set up barriers to settlement that had no compensating advantages. This was particularly true of the cases in which companies failed to complete the roads and in which land withdrawn from entry at the time the railroad filed its orginal plans remained unavailable for settlement through the uncertainties and long delays of the process of forfeiture.[67]

These conflicts could have been diminished by more careful drafting and more effective administration of the laws, and a government with a stronger fiscal base and a clearer definition of intentions could have afforded to make relatively more use of the instrument of loan subsidy and less of the blunter tool of the land grants. Yet this would have been a great deal to ask of the governments of the day, and a major reason currently given for acquiescence in the Credit Mobilier arrangements was that an attempt to construct the Union Pacific as a public work might have squandered even more money without getting the road built at all.[68] By contrast with earlier periods, in which advocates of improvements had accepted public or private or mixed enterprise with relative indifference, the era of national subsidy was characterized by a settled conviction that the federal government was not to be entrusted with the construction or operation of internal improvements. "The spirit and aim" of the acts, as one of their beneficiaries pointed out, "was to establish railroad communication . . . without the direct agency of the nation." [69] The performance of the period must be understood in terms of this limitation.

THE EMULATION OF STATES AND CITIES 1861-1890

. . . we can stand a pretty good 'steal' if we can get railroads in the State.

Fayette CHRONICLE, *1871*

Resolved: That Johnson County donate half a million dollars rather than this Rail Road should be made twenty miles east or west of us.

A public meeting in North Liberty, Iowa, 1865

6

THE STATE PROGRAMS OF THE RECONSTRUCTION SOUTH

"On the 22nd day of April 1861 the board of public works, with a feeling of the most painful regret, made an order for the suspension of work on the Covington and Ohio railroad." [1] In Virginia, as in the other states of the South, the outbreak of the Civil War brought about an almost complete cessation of action under the programs of public aid. A westerly extension of the Pensacola and Georgia Railroad, under the existing Florida system of subsidy, and the purchase by North Carolina of stock in the Chatham Railroad in 1862, were among the very few exceptions. The constitution of the Confederate States of America, reflecting what had been the prevailing view of the South toward federal action, contained a prohibition against giving aid to private enterprise. Yet the improvement of the railway network was vital to military operations, and in spite of opposition on constitutional grounds the Confederate Congress authorized as war measures loans to four railroads in the effort to fill strategic gaps in the major lines of communication—$1,000,000 for a connection between Danville, Virginia, and Greensboro, North Carolina, to complete a through line in the Piedmont region; $150,000 to the Alabama and Mississippi Rivers Railroad to fill in gaps in the line across central western Alabama; more than $1,000,000 for a connection between Rome, Georgia, and a point in northeastern Alabama; and $1,500,000 for a railroad that would join Louisiana and Texas. The smallest of the operations, that in central Alabama, was completed under direct army control before the end of 1862 —in time to be of considerable service to the military effort of the Confederacy. The more important link of the Piedmont Railroad, which had no difficulty in raising money but great difficulty in obtaining iron and labor, was not finished until 1864. On the Rome connection the work did not reach the stage of track-laying. New Orleans was captured before work was even begun on the road that was to link Texas to the rest of the Confederacy.[2]

In the North, the business of railroad promotion continued during the conflict though at a much lesser rate than in the preceding decade. In the renewed wave of railroad building that followed the end of the war, the practice of public aid was revived or expanded in many parts of the country, not only in the South and in the frontier states of the West but also in some of the richer and more fully developed states of the East and Middle West. There was, however, one marked regional difference in the nature of the programs. In the North and the Far West, railroad aid in the post-bellum period was, with few exceptions, granted by local authorities rather than by the state governments. On the other hand, all but one of the states that had formed the southern Confederacy gave state aid to railroads in the period of Reconstruction, and it is to these programs that the discussion must now turn.

There were several reasons for expecting an extensive resort to public assistance in the post-bellum South. Its railroad network was less complete than that of the North before the war, and suffered much more during the conflict. There appeared, then, much more to be done, even though the federal government at the close of the war restored much of the physical damage caused by its armies. There had been in the region, moreover, a strong tradition of public and especially of state action; the various waves of revulsion against internal improvements had affected the South less strongly than the North. As a result, no Southern state entered the period with constitutional provisions making state action impossible.[3] Finally, there were factors in the political situation which were likely to make the legislatures more vulnerable to the persuasions of the railroad promoters. Under the provisions of the legislation providing for "reconstruction" and reentry into the union, large numbers of the leading citizens of the area were debarred from voting or holding office because of their participation in the conflict at the same time that these privileges were conferred on the masses of newly-emancipated Negroes. In each of the ex-Confederate states, control of the government was held, for varying periods of time, by a "Radical" or "Radical Republican" regime composed of adventurers from the North, their allies among the Southern whites, and minorities of legislators and officeholders chosen from among the ranks of the freedmen.

Accounts of public aid in the post-bellum South commonly emphasize this last factor and indeed often seem to take the form of a drama in two acts. In the first, Radical Republican "carpetbaggers" and "scalawags"—to use their opponents' terms—join with the newly enfranchised Negroes in an orgy of extravagance, bad judgment, and corruption. In the second act, Democratic or Conservative "Redeemers"—to use their own term—throw out the Radicals, restore "white supremacy," and bring Southern governments back to ways of honesty and financial prudence. The difference between parties was in this case great enough to claim first attention in an analysis of public policy toward improvements. Yet a detailed comparison between the course of legislation and the political regimes in power shows that the two acts of the drama cannot in all cases be so sharply distinguished.[4]

In four out of the eleven states, the history of internal improvements conforms more or less closely to the pattern of Radical orgy and Conservative revulsion. This was conspicuously the case in Georgia: its post-war record was in marked contrast to its ante-bellum policy which had placed major emphasis on a single successful and highly strategic improvement, the Atlantic and Western. Yet the change began before Reconstruction was complete. The Conservative legislature of 1866 appropriated $1,500,000 for repair of the Atlantic and Western, but, in addition, it passed measures offering assistance to a number of other railroads. The Governor, insisting that the old policy of concentrated effort was still the wisest and safest, vetoed all these bills except that for the Macon and Brunswick. But the legislature of 1868–69, which had expelled its Negro members and in which control was divided between Republicans and Democrats, broke decisively with precedent and adopted legislation offering state guarantees, at rates from $8,000 to $15,000 per mile, to the bonds of seven railroads. This was still modest, however, compared to the record of 1870. In this one year of full Reconstruction, a loan of $1,800,000 was made to the Brunswick and Albany Railroad, and the legislature voted endorsements for the bonds of no less than twenty-seven railroads, at rates of $12,000 or $15,000 per mile, and in two cases increased the amounts previously authorized. During the same year, the Governor endorsed certain of these railroad bonds to an amount of over $6,500,000, of which almost $3,000,000 was later repudiated by the state on the ground that the companies had not complied with the requirements of bona fide investment and construction. An investigating committee appointed after the Democrats regained control noted that a close business associate of the Governor's was the president of the three roads on which "the State's trust" was thus abused; it also charged that the measures themselves had been "pushed through this bastard Legislature, by the infernal force of *gold*." [5]

In the case of the state road, it does not require the exclamatory rhetoric of a partisan committee report to demonstrate corruption.[6] The financial record speaks for itself. On the eve of the war, the net earnings of the Western and Atlantic, in which Georgians had taken such understandable pride, had reached $450,000 a year. It had been restored to good condition and through most of the year 1869 paid $25,000 a month into the State Treasury. Twelve months later, private interests contested eagerly for the privilege of leasing the road at a monthly rental at the same figure. Yet under the Radical administration in the year 1870, the Western and Atlantic incurred and charged to the state a deficit of at least $500,000. At the end of this remarkable year, the road was leased to a private company. There were two principal competing groups, one closer to the Governor, the other supported by Northern railroad and political interests, including Tom Scott of the Pennsylvania and the Texas and Pacific. The rivals managed to reach an agreement on the morning of the bidding, obtained the lease at the minimum rate provided by the law, and made a substantial profit over the twenty years of its operation.

The willingness of the Radicals to negotiate the lease is partly explained by a recognition of the fact that their period of power was about to come to an end. When the Redeemers came into office in the following year, they reacted promptly against Radical recklessness in railroad finance. The flood of measures for state aid ceased abruptly, although some twenty statutes authorizing assistance by local authorities were passed in the years from 1871 to 1877. The legislature of 1872 condemned the manner in which the lease of the state road had been obtained but declared that the contract itself was "a most advantageous one, as it secures a fair sum certain for the treasury, and removes the road, with its business complications, from the politics of Georgia." [7] In the same year the Democrats began the process of repudiating what they considered the fraudulent commitments of the previous regime. In 1874, the legislature repealed all earlier provisions for bond endorsement, excepting the few cases in which a vested right had been established by actual investment and construction. Delegates to the Convention of 1877 declared that one of its main objects was to put an end to railroad subsidies, and the constitution as adopted prohibited both state and local aid.

Tennessee was never in the full sense a Reconstruction state, since it was restored to the Union in 1866 before Congress had laid down the rules for reorganization. Nevertheless, its history provides one of the clearest contrasts between Republican and Democratic policies toward railroad aid. The Radicals in power after the end of the war inherited one of the most extensive of the ante-bellum systems of state aid and some 1,300 miles of railroad badly in need of rehabilitation. To assist in this process, the companies previously aided were excused from interest payments on their state bonds until 1870, and new bonds to the amount of nearly $3,500,000 were voted in January, 1866. Additional legislation before the end of the year provided aid in no less than twenty-six cases, mainly for new construction. Several other measures were adopted in 1867, and in December of that year the so-called "Omnibus Bill" granted aid to fifteen railroads. A similar measure was defeated in the following year, in part because of the vigorous opposition of the Republican state controller; but individual laws providing for assistance to seven roads were enacted in 1868 and 1869.

Under the Radical regime, the companies received a total of nearly $14,000,000 in state bonds. In violation of the provisions of both the prewar and the postwar acts, the bonds were generally disposed of by the companies at less than their par value, and the requirements of bona fide private subscriptions and of prior construction were frequently evaded. In one of the more extreme cases, the Tennessee and Pacific Railroad had up to 1870 received $1,185,000 in state bonds and $600,000 in county bonds as compared with paid-up private subscriptions of less than $17,000.[8] The majority of a committee of the Democratic legislature of 1879 looked back on all this as "a carnival of revelry and corruption" and described the Omnibus Bill of 1868 as an "assault on the Treasury" that had been perpetrated with the aid of recruits gathered "from the pulpit to the bagnio." A minority report conceded that there was "strong presumptive evidence of corruption." [9]

There was no doubt of the reaction when the Democrats came into power in December, 1869. They repealed the state aid measure even before the end of the year, and the constitutional convention of 1870 adopted a provision forbidding future assistance. This was one of the sharpest and most immediate illustrations of revulsion against internal improvements. Yet the earlier Radical measures for railroad aid seem to have had very general popular support, and a Republican paper was able to demonstrate that about nine tenths of the bonds issued had gone into the hands of Democratic railroad presidents.[10] It should also be noted that the revulsion did not extend to action by local authorities, which had been authorized by the Radicals in some twenty cases. The same legislature which repealed the state aid program adopted ten measures authorizing local aid, and in 1870 the Democrats passed a general law liberalizing the conditions for such assistance and making it unnecessary to come to the legislature for a special law in each instance.

In Alabama, the contrast between Conservative and Radical policy was somewhat less sharp. The pre-Reconstruction legislature of 1867, convinced that building railroads was the way to make "the hills and valleys of Alabama . . . vocal with industry," [11] adopted a general law offering state endorsement of railroad bonds at the rate of $12,000 per mile. This provided the basis on which the Radical regime built a still more liberal program of assistance. The first Republican legislature, meeting the following year, amended the act by easing its conditions and raising the rate to $16,000 per mile. It also adopted a general measure providing for local aid. The next legislature authorized a number of additional endorsements and raised the rate to $22,000 per mile for one road through the mountains. It also voted a loan of $2,000,000 in state bonds to the Alabama and Chattanooga. By September 30, 1871, endorsements under these various acts had reached $13,120,000, and $2,300,000 had been issued in state bonds.

The stories of lobbying and corruption center on the session of 1870–71. One legislator later testified that he had received a bribe of $60 but believed that his more sophisticated colleagues had collected several hundred dollars each. The Radical Governor admitted issuing bonds and signing endorsements in violation of the conditions requiring prior construction; in one case he made endorsements for a greater mileage than the road was authorized to build. The Democrat who succeeded him, in a partial hiatus of Radical rule, was himself accused by a committee of the Democratic House of illegal endorsements and of laxity in keeping the records of railroad aid.

No new state aid was authorized after 1870. A Republican administration acquired the Alabama and Chattanooga Railroad in bankruptcy proceedings and sold it to private interests. In 1873 a measure was adopted to substitute smaller fixed obligations for larger contingent ones by offering the railroads $4,000 per mile in state bonds for the return of $16,000 per mile in endorsements. Under this act, more than $5,000,000 in endorsed bonds were exchanged. Contemporary opinion on the measure was sharply divided. Though the act was carried mainly by Republican votes, the official organ of the party denounced it as an iniquitous scheme of the rich corporations who "put up a poor mouth . . . to suck away the life blood of Alabama." On the other hand, a Democratic paper declared that the act showed financial acumen "worthy . . . of the genius of a Neckar (*sic*)" since it would at the same time restore the public credit and stimulate internal improvement.[12]

The Conservatives came into power by the elections of 1874, and in 1875 they adopted constitutional provisions forbidding both local aid and the loan of the state's credit. Revulsion thus followed "Redemption." It should be noted, however, that the Conservatives shared much of the early enthusiasm for railroad aid and that the Radicals took some steps toward retrenchment before their final defeat.

Somewhat the same statement may be made in the case of Arkansas. Here also the Democrats, as one of their newspapers put it, "set on foot this spirit of improvement before the cormorants came." [13] The Conservative legislature of 1867 enacted a general law providing for aid in state bonds at the rate of $10,000 per mile and also adopted six measures for local aid. The Radicals came into power the following year under a constitution prohibiting state aid except by vote of the people; but their first legislature promptly submitted to referendum a liberal law providing aid at $10,000 per mile for roads entitled to federal land grants and $15,000 a mile for all others. After a lively campaign, the measure was carried by an overwhelming vote cast on ballots which read simply "For Railroads" or "Against Railroads." A swarm of promoters came forward to take advantage of the new law, some of them evidently counting on their political influence and many of them new men unlike the old-time Democratic railroad presidents of Tennessee. No less than eighty-six companies received charters, and the Republican Governor was accused of irregularities in the issuance of bonds. In the end, however, only five of the many corporations, after some consolidations of smaller companies, actually received state bonds. The total issued was $5,350,000. The last Radical legislature, meeting in 1873, adopted a general law providing for local aid and debated a bill to permit the railroad companies to give to the state preferred stock in exchange for the release of their mortgage obligations. It was argued that such a subordination of the public claim—like that of the United States on the Union Pacific—was essential in order to obtain private capital for necessary expansion; but the measure failed of enactment after bitter controversy. In 1874 the Conservatives returned to power, and before the year was over adopted constitutional provisions prohibiting both state loans and local aid.

As against these four instances of Radical "orgy" and Conservative revulsion, there were four other states in which the first half of the story was similar but the second quite dissimilar. In the case of Florida, charges of fraud in the Radical period can be substantiated from the records of the United States Supreme Court.[14] The principal recipients of aid were the Florida Central and the Jacksonville, Pensacola and Mobile, ante-bellum roads which had been taken over by the state after default and had come under the control of "The Prince of Carpetbaggers," Milton S. Littlefield, and his North Carolinian ally, George W. Swepson.[15] The latter had bought one of these roads from the Republican administration, paying for it partly with a worthless check and partly with bonds of the companies themselves, which he had bought at low prices with funds embezzled from a North Carolina road. In spite of a provision in the Reconstruction constitution which prohibited taxation for the benefit of a corporation, the legislature of 1869–70 adopted measures offering aid by state bonds or endorsement to six railroads at rates from $10,000 to $16,000 per mile. Under this legislation, the Governor promptly issued $4,000,000 in bonds to the two Littlefield-Swepson companies. The bonds were sold at a fraction of their face value, and an even smaller fraction of the proceeds was devoted to the building of the roads.

What distinguishes the Florida case from those discussed earlier is the fact that revulsion took place some years before the state was "redeemed" by the Democrats in 1877. Most of the bonds provided for by the laws of 1869 and 1870 were never issued. Early in 1873, an act was passed repealing all measures providing for aid by endorsement or by state bonds. Finally, a constitutional amendment was adopted in 1875 forbidding railroad aid by local authorities and either stock subscription or the loan of credit by the state. These actions correspond almost exactly with those taken in Georgia in 1874 and 1877, but the difference is that in Florida they were carried out not by Conservative Democrats but by Radical Republicans.

Financial aid does not comprise the full Florida story, and both Radical and Conservative regimes found ways of aiding railroad construction by the use of other assets. The state had "swamp and overflowed lands" at its disposal and also land granted by the federal government in 1850 for purposes of internal improvements. The pre-Reconstruction government of 1866 made one railroad land grant, and the Radicals added several in 1870. In 1874, after repealing the measures for financial aid, the Republicans adopted a general law setting conditions for grants or sales of these lands. The Conservative legislature of 1879 continued the policy with grants of land to six railroads.

North Carolina's experience was like that of Florida in two respects. Both were victimized by the same pair of promoters who used "the funds of one state as a means of robbing another, while seeming to serve both states in patriotic purposes." [16] In each, also, a sharp reaction against reckless railroad aid took place before Redemption. But in North Carolina orgy and revulsion were compressed into shorter compass within the briefer period of Radical rule.

North Carolina Republicans inherited a program and tradition of state encouragement to internal improvements. At the close of the war, the state held a two-thirds interest in three North Carolina railroads and had investments in several others. The pre-Reconstruction legislature of 1866–67 renewed the antebellum measure requiring the state to bear two thirds of the cost of building the Western North Carolina, and it aided two railroad companies by transferring to them securities held by the state in other internal improvement enterprises. It also adopted a number of measures for local aid.

When the Republicans came into power, the pace of subsidy

was sharply accelerated. They were so anxious to aid the railroads that the Constitutional Convention of 1868, without waiting for the legislature to meet, adopted ordinances offering a $1,000,000 endorsement to one company, $1,200,000 in state bonds to a second, and bonds to a third at the rate of $10,000 a mile. When the General Assembly met on July 4, it received a message from the Governor urging the creation of "a vigorous and well-directed system of internal improvements, from the seashore to the Tennessee line." [17] Before it finished the sessions, the number of Radical measures offering state aid had risen to sixteen. By the time the next Assembly met in the fall of 1869, bonds had been issued under the new legislation to the amount of $12,600,000, and about half of these had gone to the Western Division of the Western North Carolina.

The legislature of 1869–70, though still Republican, was in a different mood. In February it ordered the companies to return all state bonds that remained unsold. In March it passed a bill repealing all railroad appropriations made by the preceding legislature. The reasons for this reversal lay partly in the price of the state's bonds, which had fallen in the panic of 1869 from fifty or sixty cents on the dollar to twenty-five cents or less, and partly in the findings of a Senate committee on what the companies had done with the bonds they had received. On most of the roads, they had either been applied to construction purposes, subject in a number of cases to deductions to reward General Littlefield for his services in lobbying, or had not yet been sold. More than $4,000,000 of bonds were returned under the February act. But the major case of the Western Division, whose presidents were successively Swepson and Littlefield, was a different story. They had spent almost nothing on building the road, they had raised almost no private capital for it, and they had used almost $1,000,000 of North Carolina bonds in the Florida speculation which has already been described.[18]

The Conservatives came into power in the elections of the same year and continued the policy of revulsion. They made further investigation of the railroad scandals and made prolonged and partially successful efforts to obtain restitution. The legislature of 1871–72 attempted to reduce the state's railroad obligations by an act offering to relinquish the state's holding in stocks which would return the state's bonds. Yet the North Carolina Conservatives were by no means disposed to sell at a sacrifice the entire railroad investments of the state. In 1877 the state agreed to pay an assessment of $1.50 per share on its stock in the Atlantic and North Carolina and in 1879 to aid another road by purchasing its first mortgage bonds. Of greater interest was the action taken in the case of the Western North Carolina Railroad, which the state bought in 1875 for $850,000 and transformed into a mixed enterprise, retaining in its own hands three quarters of the stock.

Local aid to encourage railroad construction, which had been employed both before and during Radical rule, was relied on even more extensively after the Redeemers came into power. Their first legislature, meeting at the time of greatest disillusionment over the ruined credit of the state, enacted no less than twenty-one measures providing for local assistance. The policy was maintained throughout the decade, with ten such acts adopted in 1879. They contained various provisions to ensure actual construction, including, for example, the requirement that the subscription should not be made until the road was built and "cars running" to a given point. Most measures called for bond issues, in one case limited in amount to $10 per inhabitant; but a number permitted subscriptions in "lands, timber, work or money." As these acts indicate, the reaction in North Carolina against the excesses of state subsidy led not to the abandonment of public aid but to reliance on more modest expedients for stretching the limited resources of an agricultural area for the improvement of its transportation facilities.

The case of South Carolina displays the same pattern of state aid by the pre-Reconstruction legislature, of its intensification under the Radicals, and of revulsion before the Democrats returned to power. The principal controversies were not over the amount of aid during Reconstruction, which was relatively small, but over the terms on which the Radical regime disposed of the state's interest in certain railroads.

In 1866, the Columbia and Augusta was voted bond endorsement at the modest rate of $6,500 per mile, and the Greenville and Columbia was granted a $600,000 addition to earlier endorsements. In 1868, the first year of Radical rule, the chief claimant for assistance was the Blue Ridge Railroad. In spite of its antebellum failure to cross the mountains, the president of the company still insisted that it would provide "the cheapest and most direct line from the West to the Atlantic seaboard." [19] The legislature responded by voting $20,000 in cash for urgent repairs, $1,000,000 of endorsement of its existing bonds, and $3,000,000 of endorsement for new bonds. In 1871 the state agreed to defer its own lien on the company to permit it to seek first-mortgage financing elsewhere, and in the same year the state's stock interest in the road was sold at $1.00 a share to a "ring" which included officials of the state administration. Because of the depreciation of the state's credit, the endorsed bonds remained unsalable. In 1872, the Radical legislature, over the veto of the Radical governor, passed a measure offering the company $1,800,000 in state scrip in exchange for the $4,000,000 in endorsements. The story of the Greenville and Columbia, which obtained additional aid in 1869, is a similar one. The state's stock was sold to the "ring" in 1870; and this road also, on consolidation with the Blue Ridge, obtained deferment of the state's lien. The combined company failed to prosper, and the great project of the western connection never advanced beyond the tunnel in Stump House Mountain.

Reaction followed quickly. As early as November, 1869, the Senate endorsed a committee report which opposed the granting of further pecuniary aid to any railroad scheme. Committees of the two following legislatures criticized the handling of state finances and investigated charges of bribery to secure the passage of railroad measures.[20] In January, 1873, the legislature, still under Republican control, ratified a constitutional amendment forbidding "the loan of the credit of the state by guarantee, endorsement or otherwise" without the approval of two thirds of the qualified voters. The next legislature began the process of scaling down the debt, much as the Conservatives were doing in other states at the same time. Meanwhile, with state aid abandoned, Radical legislatures attempted to meet the continuing demand for encouragement to railroad construction by measures for local assistance. The Democrats continued the same policy after they returned to power in 1877.

In Louisiana, state financial aid to railroads was extended during five years of Radical rule, but the policy was reversed long before the Conservatives returned to power. In the twelve measures passed from 1868 through 1872, the most common form of aid was by bond endorsement at the rate of $12,500 per mile to be covered by a second mortgage. A number of the projects failed to materialize, and the principal aid actually extended was to the New Orleans, Mobile and Texas Railroad Company, which was to make the much desired connection with Texas. Acts of 1869, 1870, and 1871 provided for aid in the successive forms of endorsement, bond subsidy, and stock subscription, under which the company received something over $4,000,000. With this assistance it completed some seventy miles of road but abandoned operations far short of the Texas border.

The railroad measures aroused violent controversy. Accusations of bribery to secure their passage were matched by accusations of the use of money by competing shipping interests to prevent their adoption. The alignments did not conform completely to party lines. A Republican Governor vetoed several subsidy bills and warned the legislature against "schemes of plunder" put forward by "well-dressed gentlemen claiming to be representatives of the most respectable of our people." On the other hand, Democratic editors and legislators were among those who spoke most eloquently of the need to reach out to the great state of Texas, that "Canaan of promise . . . at our door."[21]

After 1872, the Republican legislatures passed no further acts offering state financial aid. A repeal measure was adopted in 1874. As early as 1870, an amendment was adopted limiting the public debt of the state to $25,000,000, and the limit was reduced to $15,000,000 by action of the 1874 legislature. Meanwhile, the state continued to grant to new railroads free rights of way and use of materials in the state lands, and made occasional use of the method of local aid.

In Louisiana, Redemption was marked not by revulsion against railroad aid but by a renewal of interest in railroad projects. The Democrats enacted two ways-and-materials measures in 1877, even before the withdrawal of federal troops made their government secure, and added three others in the following year. They repealed one Radical bill for local aid but enacted two of their own. Connection with Texas was again the great object, and in 1878 the legislature voted to lend $2,000,000 in state bonds to the Texas and Pacific. The effort failed, somewhat ironically, when the Democratic Governor refused to sign the bonds on the ground that they would put the state debt beyond the limit set by the Republicans' constitution.

There remain three of the ex-Confederate states which conform to neither of the two patterns. Of these, however, Mississippi was less of an exception than has sometimes been supposed. State aid was apparently barred by a constitutional provision adopted by the so-called "Black and Tan Convention" of 1868, before full Reconstruction, prohibiting the loan of the state's credit. Yet the Republican legislatures, in power from 1870 through 1875, found a variety of means of encouraging railroad development including the granting of right of way and materials on state land, the sale of land at low prices, and the transfer of securities previously held in other railroads. In 1871 a controversial measure was adopted appropriating $4,000 per mile for railroad loans out of actual or anticipated balances in the State Treasury, and in 1873 the amounts per mile were increased for two railroads. The first mileage law received no Democratic votes in the legislature, but much of the Democratic press reached the conclusion that "its intrinsic good" outweighed its "evil parentage." [22] When the Redeemers came into power in 1876, they rejected a proposal to repeal the law granting $8,000 a mile to the Vicksburg and Memphis. What lavishness there was, therefore, was Republican. The total of state aid, however, was less than in most other states, and the reaction against it was correspondingly moderate. On the other hand, both regimes made vigorous use of the ante-bellum policy of local aid, the Radicals enacting a general law and nearly fifty special laws for the purpose and the Conservatives adding sixty more before the end of the eighties.

Virginia, which did not experience Radical Reconstruction, was the only state which provided a complete exception to the practice of state financial aid. As the result of its vigorous ante-bellum program of assistance to internal improvements, the state at the end of the war held a majority of stock in most of the Virginia railroads and was the sole owner of the Covington and Ohio and several others. The roads were in poor physical condition, like those of the rest of the South, but Virginia chose a different solution from that of the other states. The constitution adopted in 1869, before the Conservatives were fully restored to power, prohibited state stock subscription and grants of credit. The decisive step was taken in 1871, when the legislature voted to sell its railroad securities and apply the proceeds toward the payment of the state's debts, which had been incurred largely for railroad purposes. Virginia's trunk line to the West, on which work had stopped at the outbreak of war, was transferred to private management under the new name of Chesapeake and Ohio. With aid from the city of Richmond and from counties in Virginia and West

Virginia, the C. & O. finally reached the Ohio River in 1873. Railroad leaders took an active part in state politics while these decisions were being made, and in the case of Virginia charges of undue subservience to railroad interest have been made not against Radicals for lavish subsidy but against Conservatives for selling railroad assets at too great a loss and without making adequate provision for future regulation.[23] On the other hand, the Virginia legislature gave an even freer rein than most states to local assistance, adopting a general law in 1867 and some 140 special authorizations in the period 1866–89.

The case of Texas differed from all the others in methods of aid, in their timing, and in their relationships to political alignments. The principal distinguishing factors were the vast amount of public land remaining at the disposal of the state and the position of Texas on a major transcontinental route.

In 1866 a legislature controlled by Conservatives revived the ante-bellum "policy of 1854" by enacting a law providing for land grants for all railroads built within the state to the amount of sixteen sections per mile. It also passed a general law authorizing assistance by local authorities. More surprisingly, the constitutional convention of the same year adopted a provision, which was never used, permitting state endorsement of railroad bonds at the rate of $15,000 per mile.

On the other hand, the framers of the Reconstruction constitution, which was adopted in 1869, regarded the land grants as profligate and forbade them altogether—a restriction which remained in force throughout the years of Radical rule. During this period, the greatest controversy, however, occurred over two measures providing for the donation to railroads of large issues of state bonds. Charges of corruption were freely made, and party lines were broken in the disputes. The most hotly debated measure, which offered $6,000,000 in bonds or alternatively 14,400 sections of land to two companies of what became the Texas and Pacific system, was carried in 1871 largely by Democratic votes over the second veto of a Radical governor.[24] The bonds, however, were never issued.

Redemption brought about an abrupt change but not in the conventional direction. The self-styled "Liberators of Texas," taking over control of the legislature in 1873, removed the restriction on land grants, chose the land as against the bond alternative for the Texas and Pacific subsidies, and before the year was out had passed measures offering sixteen sections or more per mile to no less than twenty-seven railroad corporations. Thirty more land grants were authorized in the two years following, and a general law providing for sixteen sections per mile was adopted in 1876. The constitution of 1875 prohibited local aid, which had been practiced both by Radicals and Conservatives, and forbade state grants of money. But the land grant policy was not abandoned until 1882, by which time the public domain was virtually exhausted and railroad grants had reached a total of more than 32,000,000 acres.

In Texas, therefore, effective state aid took the form of land rather than credit, since neither the Radicals' bonds nor the Conservatives' contemplated endorsements were ever issued. It was extended much more liberally by Conservatives than by Radicals. Most of it, moreover, was given from 1873 on, rather than before. In all these respects, this record was unique in the history of the Reconstruction South.

In the post-bellum South, there were, as we have seen, nine states whose governments, unlike those of Virginia and Texas, gave direct financial support to railroad development. By loans or guarantees or capital outlays on state roads, they committed themselves to expenditures and obligations that appear to have reached a total, at face value, of between \$60,000,000 and \$70,000,000. In terms of current impact, the figure must be reduced to allow for discounts on the sale of state bonds, and a number of the bond issues were later repudiated. By far the greater part of the financial aid was concentrated in the years 1868–72, and in most cases the period of lavish grants had come to an end before the financial crisis of 1873. After that year no new program of importance was initiated. In Georgia, Tennessee, Alabama, and Arkansas, and perhaps even in Mississippi, the change from recklessness or generosity to retrenchment coincided rather closely with the shift from Radical to Conservative rule. But in Florida, in North and South Carolina, and in Louisiana, the same reversal of policy took place while the Radicals were still in power. "Redemption" was evidently not an essential condition for the revulsion against improvements. Republican regimes showed that they also could draw the moral of retrenchment, either because of their own disillusionment with chicanery and failure, or because of the pressure of public opinion, or for the cogent reason that the fall in the price of state bonds made it impossible to support their programs.

Yet certainly the nature and origins of the programs were colored if not determined by the character of the political regimes in power. It may even be asked whether in this period state aid would have been given to Southern railroads at all in the absence of Reconstruction—that is, if we can imagine the political and economic dominance of the groups previously in power left undisturbed by federal restrictions, by the arrival of ambitious Northerners, or by the grant of political rights to the freedmen. A negative answer is perhaps suggested by the fact that the state least affected by Reconstruction, Virginia, was the only one which refrained from state assistance. Yet the circumstances of Virginia, which had one of the two largest railroad networks when the period began, were quite exceptional: there are strong reasons for believing that through most of the area the credit and resources of the state governments would have been used for the encouragement of railroad building. This policy was in line with ante-bellum tradition, and Conservative regimes in Arkansas, Alabama, Georgia, and South Carolina had begun to revive it with vigor and even liberality before the Radicals came into power. Throughout the period, moreover, the Democratic press continued to emphasize the South's backwardness in railroad construction and to draw from it the moral that every effort must be made to attract outside capital. This was illustrated in the widespread campaign for federal assistance to the Texas and Pacific and other Southern enterprises which ran counter to the "traditional creed" of the party and section; in an editor's warning against allowing Granger attitudes toward regulation to gain a foothold while it was still necessary to tempt "foreign" capital for railroad building; and in the eagerness with which Swepson and Littlefield were welcomed in Florida in the belief that they were bringing money into the state in order to develop its resources.[25] The same feeling was expressed in many comments on the state programs themselves. "The Legislature," said one Democratic organ, "may have had in view the stealings they could make out of the scheme; but we can stand a pretty big 'steal' if we can get railroads in the State."[26]

It may be assumed that a different regime would have permitted fewer "stealings" and avoided certain of the frauds which "disgraced the pages of American History."[27] But it would have been difficult for any governments in power in most of the states to have resisted the popular demand for railroad aid at the end of the sixties and difficult for any governments to have raised enough funds by taxation to maintain the states' credit without curtailing or abandoning the programs in the depression of the seventies.

If state aid had been refused, it seems clear that fewer railroads would have been built. A study of the figures of construction indicates that in most of the South, mileage built per year was greater under Radical than under Conservative regimes, and that the reverse was true in Texas, where the Conservatives gave greater largesse than the Radicals. The case of Florida, to be sure, stands out as that of a state which acquired railroad debt during Reconstruction and little railroad mileage. On the other hand, Mississippi acquired no debt and little mileage. Virginia, which gave no state aid, shows the smallest percentage increase. In Arkansas 662 miles were added under Radical rule, and the amount built under its much criticized system of state aid corresponded closely to the amount for which bonds were issued. Alabama also showed a notable increase in mileage under Reconstruction. Georgia's peak figure of 263 miles in 1871 presumably owes more to the Republican largesse of the previous year than to the restraint of the Redeemer legislature that met in the fall of 1871. In the years 1866–79, the eleven states increased their railroad mileage by some 7,000, of which more than 2,000 was in Texas. This was a substantial achievement and did much to transform the short and often disconnected lines of 1865 into a relatively complete regional system, though it did not keep pace with the rate of increase in the nation as a whole. In the case of Texas, 500 miles were built in the year 1873, when the Liberators were granting land to twenty-seven companies, and another 1,000 in the six years that followed. In the other states, however, well over half of the gain was accomplished in the short period 1870–73.[28]

Retrenchment did not build railroads, whether practiced by Conservatives or Radicals. Whatever the leakages of public funds, and in spite of the fact that some roads were built without government assistance, there was—as would be expected in a developing region—a clear and substantial correlation between railroad aid and railroad construction.

It would, however, be a mistake to attribute these differences to political causes alone. Texas and Arkansas, for example, lay across the lines of promising transcontinental routes; on the other hand, Virginia had perhaps the least need for new construction. It would also be a mistake to consider the achievements and the defects of the Southern programs without relating them to those of the rest of the country. For the nation as a whole, as well as for the South, the years preceding the panic of 1873 represented a period of boom construction and those which followed a period of relative stagnation, though the contrasts were somewhat sharper in the South. In many parts of the North, the years 1868–72 constituted a time of particularly active railroad aid, and the later seventies a period of revulsion against the policy. The sequence of enthusiasm, miscalculation, and corruption, and subsequent disillusion, can be illustrated by Northern as well as Southern examples. In both North and South, moreover, the persistent demand for railroads in areas lacking capital continued to express itself in local aid long after the abandonment of state programs. Even the drama of Reconstruction did not make the Southern experience wholly unlike that of the rest of the nation.

7

"THE PATRONAGE OF LOCAL LEGISLATION"

The revulsion against government promotion of internal improvements had made considerable headway in the North before the outbreak of the Civil War. In the period that followed, public aid was not as widespread as in the states of the Reconstruction South. In a number of areas in the North, extraordinarily large programs of local aid coincided in time with the greatest burst of state activity in the South. Here also there was a reaction in the early eighteen-seventies, yet in various parts of the country railroad companies continued to seek and receive "The Patronage of Local Legislation" for almost as long as railroads continued to be built.

The most successful of the state-owned canals—the Erie, the Sault Ste. Marie, and the Illinois and Michigan—all remained in operation. The last of these, which had been held by trustees for the bondholders, came back into the full possession of the state of Illinois in 1871 with all debts paid. By that time the state had already begun improvements on the lower part of the Illinois River. These were completed in 1877 at a cost of nearly $750,000 and turned over to the Canal Commissioners for operation and the collection of tolls. Net income reached its peak in the late eighteen-sixties; from 1879 on, however, there were operating deficits instead. Tonnage rose to a maximum of over a million tons in 1882 and then gradually declined to less than half a million by the end of the century.[1]

The Soo and the Erie both became toll-free waterways. In the former case, the change was accomplished by transfer to federal ownership in 1881, and it was the national government rather than the state of Michigan that bore the costs of enlargement and modernization to serve an increasing volume of traffic.[2] In New York the Erie Canal earned a good return in the eighteen-sixties, but during the next decade its net income was reduced as the result of lower tolls and increased railroad competition. The southern laterals, which were clearly unprofitable, were abandoned in 1878. The Erie, Champlain, and Oswego Canals were retained in state operation. An amendment establishing the policy of "free canals" was adopted in 1882, partly on the argument that continued canal competition would keep down railroad rates, and collection of tolls ceased at the end of the year. Minor improvements of the canals continued to be made, and early in the twentieth century the state transformed the system into the modern Barge Canal at a cost of some $130,000,000.[3]

In railroad development the Northern state governments of the period played only a limited role, either on the Western frontier or in the more settled areas with established networks of transportation. By 1861 state participation in the promotion of internal improvements had been forbidden by constitutional provisions in Maine, New York, Pennsylvania, and Maryland in the East, in all the states of the Old Northwest including Minnesota, in neighboring Iowa and Kentucky, and the new state of Kansas, and in California and Oregon in the Far West. During the eighteen-sixties, West Virginia, Nevada, and Nebraska entered the Union with similar prohibitions.[4] Since state aid had not been given to railroads in the smaller New England states or New Jersey and had been abandoned in Delaware, Missouri and Massachusetts were the only states in which state promotion of improvements was sanctioned both by law and current practice. Of these Missouri was just at the point of revulsion against its generous program of state aid to railroads. The discussion will therefore begin with the quite exceptional case of Massachusetts and then go on to the examination of the more widespread and persistent practice of local aid.

At the beginning of the Civil War decade, Massachusetts, like South Carolina, had driven a tunnel for an intersectional railroad part way into a mountainside. Massachusetts finally holed through and South Carolina did not, but the difference was not necessarily to the advantage of the richer of the two states. Work on the Hoosac Tunnel was brought to a halt in 1861 under circumstances which gave rise to a war of pamphlets and investigations. At this time the Troy and Greenfield Railroad, which had raised almost no private subscriptions, was virtually the property of its engineer-contractor Herman Haupt, formerly chief engineer of the Pennsylvania Railroad. Haupt had risked his own capital—and reputation—on the success of the undertaking to which he had been attracted, according to his own account, by its "magnitude, peculiar character, and engineering difficulties." [5] It was already clear that the magnitude and difficulty of the work had been greatly underestimated. The small contributions of the towns and about half of the state's conditional loan of $2,000,000 had been spent. There still remained four miles of the five-mile tunnel to be dug, as well as some railroad-building at the eastern end. In this situation, further payment on the state contribution was suspended, Haupt was dismissed, and the state took over possession. Strong arguments were put forward for the abandonment of the undertaking—one pamphleteer even insisting that Massachusetts could well afford to pay $1,000,000 to anyone who would take the project off its hands [6]—but the same political pressure from the northern counties that had been responsible for the initiation of the work remained strong enough to ensure its continuance. Three Commissioners, themselves successful railroad men, were appointed to take charge of the construction as a state work. Haupt predicted that "their education" in the Hoosac Mountain academy would cost the state more than his own had done; [7] in fact, they spent over $3,000,000 of its money without completing the work. Finally, after the state had turned over construction to a firm of private contractors at an additional cost of nearly $5,000,000 and after an effective system of power drilling had at last been applied, "the Great Bore" was finally holed through in 1873.

Completion of the tunnel did not bring to an end the state's connection with the enterprise or its expenditures on it. It was decided to maintain the state's ownership of the railroad, partly in the hope that it would serve as a yardstick to measure the reasonableness of the rates charged by the competing Boston and Albany. The road, however, was operated under an anomalous "toll-gate system" under which the connecting railroads ran their trains over it and paid a part of the proceeds to the state. By the mid-eighties it was evident that this type of partial state operation was a failure; in the prevailing climate of opinion the alternative of full state operation was not to be considered. In 1887 the Troy and Greenfield was sold to the Fitchburg Railroad which also bought the western connections which gave it a through route from Boston to Troy. The state received $5,000,000 in the Fitchburg's bonds and $5,000,000 of its common stock. Though the stock had little value at the time, it was replaced in a later consolidation by bonds of the Boston and Maine. Since the state had borrowed more than $14,000,000 for its investment in the Troy and Greenfield, its final loss on capital account may be thought of as about $5,000,000; but a calculation taking account of net interest payments would bring the total figure to nearly $9,500,000.[8]

By comparison with the expenditures that other states had made before the war for developmental improvements, the sum does not appear extraordinarily large. Moreover, Virginia's Blue Ridge Railroad, with its tunnel built by Charles Crozet, had already established the precedent of using the agency of the state to overcome the mountain obstacle on a through line that was not itself a state enterprise. But Massachusetts had neither Board of Public

Works nor State Engineer; and the processes by which the tunnel railroad was built were less the results of rational calculation and firm execution than reflections of the shifting political battle between the tunnel bloc of the northern counties and the supporters of the Boston and Albany. As finally achieved, the railroad fulfilled only a small part of the claims that had been made for it. It brought local benefits to the towns that had agitated so insistently for the state's support; and it provided a through route between Boston and the West with lower grades than those of the Boston and Albany, though this operating advantage was for the most part canceled out by much higher fixed costs. Under these conditions, the road became a major route for freight but not for passenger traffic. It was to render particularly important service as a freight carrier during the two World Wars. Yet if the balance had been struck at the end of the nineteenth century, at the time when the state ceased to be a stockholder in the Fitchburg, it would have been difficult to argue that the Commonwealth had achieved a developmental gain commensurate with its evident financial loss.

The same sort of political pressure for aid to sections of the state which possessed less natural advantages was also reflected in the loans totaling $1,300,000 to two railroads in the Berkshires. Similar local ambitions were expressed in a new wave of town aid, which was in part regularized by a general act in 1870. Favorable votes were cast in more than fifty town elections, but the total granted did not greatly exceed $2,000,000. Springfield, already a major point on the Boston and Albany, voted aid of over $600,000 for two other railroads. Northampton's $300,000 was the largest of the twelve town subscriptions to a third railroad between Boston and the Connecticut River, the struggling and unprofitable Massachusetts Central. Other smaller enterprises represented the efforts of towns and villages by-passed by the main lines to ensure their survival in the railroad age by the building of spur or branch lines.[9]

The remaining case of public aid in Massachusetts can find no such explanation in the local interest of the less favored communities. The Boston, Hartford and Erie, in which the Commonwealth of Massachusetts became at one time the principal stockholder, was to do most of its building outside the state. The object was to create an intersectional railroad between the areas exploited by the Boston and Albany and the shore-line route. Such a road, according to its promoters, would have the special advantage of giving Boston a connection with the West and the Pennsylvania coal fields that would be independent of New York City. The chosen route was from Boston and Providence to Hartford and then to Fishkill, New York, to make the connection with the Erie's terminus at Newburgh on the other side of the Hudson. In an earlier form the project had received support from the cities of Providence and Hartford before the war. The new company was chartered in 1863 and came under control of the Erie Railroad interests in 1866. Its securities were largely unsalable; but in the following year the Massachusetts General Assembly, acting on the unanimous recommendation of the House Committee on Railways and Canals, came to the rescue by authorizing a loan of $3,000,000, to be issued on the security of $4,000,000 in the first mortgage bonds of the railroad. Additional aid was voted in 1869. The state scrip was to be issued only on assurances that the company had funds enough to fill in the gaps for a completed line from Boston to Fishkill and—under the second act—on the basis of matching expenditures from private sources. By 1870, it was glaringly evident that neither condition had been fulfilled: issuance of the state scrip was ended after its total had reached $3,600,000. Only a governor's veto, however, prevented extension of a further loan for the completion of the road. The remainder of the story is a series of bankruptcies and reorganizations. When the company emerged from one receivership in 1875 as the New York and New England, the state was its largest stockholder. Part of this stock was later sold on the open market, but the greater part was exchanged, on a two to one basis, for second mortgage bonds of the company. These in turn were disposed of during a second receivership in 1885, and the state's financial interest in the enterprise was terminated. Though the final price was held up by the somewhat unexpected competition of two rival groups of speculators, the state's net outlay on principal and interest amounted to more than $8,000,000. Meanwhile, completion of the road to Peekskill, which was finally accomplished in 1881, had served only to demonstrate that the hopes for a great intersectional traffic were quite unfounded.[10]

The post-bellum experience of Massachusetts appears difficult to understand. If public aid by this time was mainly to be thought of as an expedient of areas lacking capital, as Poor and others insisted, it seems strange to find so much of it in a state whose capital, Boston, was itself one of the principal markets for private investment in Western railroads. It might, however, be argued that so wealthy a state could afford to contribute to the development of its less profitable areas; and part of the Commonwealth's expenditures, however ineffectively administered, may be thought of in these terms. Nor is there difficulty in understanding the efforts of the towns to obtain railroad connections by means of public aid. On the other hand, the state's entanglement with the Boston, Hartford and Erie is harder to comprehend. From this experience, which was marked by fraud as well as folly on the part of the promoters and by laxness as well as bad judgment on the part of legislature and administration, a House committee drew the conclusion "that, from the nature of the case, the Commonwealth must be the losing party in any business transactions in which it engages with private citizens." [11] It is interesting that this should have been said, not of a frontier state in the boom of the thirties, and not of some Southern state in the grip of Reconstruction, but of the seasoned Commonwealth of Massachusetts in 1878.

In other Northern states, local authorities played the principal role in public assistance to railroads, and the state governments took minor parts or none at all. In a few cases local aid represented the far-flung designs of great commercial cities, but in most cases participation was widely spread among large numbers of governmental units, including cities, towns, counties, and sometimes even the smallest villages and townships. The most notable programs of railroad promotion by great cities were those of Baltimore, which continued its ante-bellum practice of investment in mixed enterprise, and of Cincinnati, which took the bold and somewhat unexpected step of municipal railroad building.

In Baltimore, the $6,000,000 of railroad aid given between 1864 and 1886 was about half the amount given before the war: nothing in the post-bellum program matched in developmental importance the earlier aid that had done so much to ensure the success of the Baltimore and Ohio.[12] The affairs of this railroad continued to be matters of municipal concern. At the beginning of the sixties, the City Council lobbied successfully against a measure proposed by the B. & O.'s new president, John W. Garrett, that would have given private interests a majority on the board of directors even while the state and city held a majority of the stock; but his object of private control was attained when the Maryland stock was sold in 1867. Meanwhile the president had already demonstrated his ascendancy by securing consent to the railroad's control of a connecting line in Ohio over the protests of the Mayor and a minority of the Council's Joint Committee on Internal Improvements, who thought the city's investments were being jeopardized because its "railroad Alexander" was ambitious to "cross the Ohio, like another Indus." [13] These controversies, however, did not prevent Garrett from appealing to the city to invest $1,000,000 in the Valley Railroad and $500,000 in the Lynchburg and Danville. These two Virginia roads were to be extensions of the B. & O. system, but its president argued that the company had relatively little to gain from them "as a mere railroad matter." The great advantage would, he said, accrue to the city in "net profits on the business that will thus be commanded."[14] In the referendum of 1869, the voters rejected the Lynchburg and Danville proposal but approved the subscription to the Valley Railroad. The road, however, never completed the southern connections that were to have "commanded" an extensive business, and the city received no return on its investment. Baltimore's prewar interests in two other branches of the B. & O. system, the Northwestern Virginia and the Pittsburgh and Connellsville, were disposed of to the parent company, in 1864 and 1875, at heavy losses. On the other hand, the city's stock in the B. & O. itself paid dividends without interruption from 1863 to 1886. The city missed the opportunity to sell out at a substantial profit when the dividends were at the rate of ten percent, but had managed to break even on capital account by the time it sold its last shares in 1890.

In this period the city's major promotional effort was devoted to the Western Maryland. The immediate purpose of the road was to tap the trade of the northwestern part of the state, which the B. & O. did not reach; but some of its projectors hoped to make it the basis of a rival through route to the West, and the President of the Baltimore and Ohio denounced the idea of building "an unnecessary and certainly unremunerative parallel line."[15] Hopes of a through road to Pittsburgh were based first on a connection with the Pittsburgh and Connellsville, in which the Western Maryland was forestalled by the B. & O., and later on a connection with the projected South Pennsylvania Railroad, which was never built. The main line of the Western Maryland was completed to Hagerstown in 1872 and was later pushed westward to the Maryland coal fields.

This limited accomplishment was made possible largely by municipal funds, and the city took over the principal share in the control of the company. Endorsements of company bonds in 1860 and 1864 and a stock subscription of $200,000 in 1866 were followed by more extensive aid in 1870 and 1872. A final loan of $1,800,-

000 to the company in 1886, to make possible refunding of its various obligations, brought the total of city aid to more than $5,500,000. On these contributions Baltimore received no financial returns until after 1890. If interest on the principal of loans and endorsed bonds is added to the city's initial outlays, its capital investment as of 1902 may be thought of as exceeding $9,000,000. The old dream of a great Western connection affected the final sale of the city's interest in that year: the Gould syndicate's bid of $8,750,000 was accepted, as against others that would have shown a net profit, in what turned out to be the unfounded hope that the purchasers would make it a part of a through system extending to the Southwest.

The bond endorsement of 1870 was given on condition that the city appoint a majority of the board of directors, and for some thirty years the Western Maryland was operated substantially as a municipal enterprise. During this period both its physical and its financial position—after unpromising beginnings—were gradually improved. It carried to Baltimore a substantial volume of local freight and coal but at heavy cost to the city's taxpayers. In this case, the inhabitants of a relatively unprosperous area owed their railroad not to their influence in a state legislature but to the ambition of a commercial city to possess a complete network of railroad connections.

Another municipal enterprise, the Cincinnati Southern Railway, was initiated under very different circumstances. The city of Cincinnati, which in the 1860 census barely maintained its position as the largest center west of the Appalachians, was clearly losing ground in competition with rival cities. In particular its commanding position in trade with the South was being taken over by Louisville as railroads came to assume dominance over river transportation. What Cincinnati needed, in the view of its commercial leaders, was a railroad to the South that would meet the threat and challenge of the Louisville and Nashville. Private interests, however, were not prepared to build at their own cost and risk over the rough terrain of eastern Kentucky and Tennessee. The familiar expedient of mixed enterprise, which had been employed in the city's earlier subscriptions to the Little Miami and other railroads, was blocked by the constitution of 1851 which forbade municipal subscriptions, loans, and donations to private corporations. An attempt to raise money by popular subscription to give capitalists a "bonus" for undertaking the venture failed on the eve of secession, and Congress did not adopt President Lincoln's proposal to build the railroad as a war measure.

In this situation a Cincinnati lawyer proposed a bold and somewhat novel solution. Noting that the constitutional "prohibition was against the city being a stockholder and not against accomplishing a public object out of her own means," [16] he suggested that the city build the railroad on its own account. A joint committee representing the City Council, the Board of Trade, and the Chamber of Commerce sponsored an enabling act which was passed by the legislature, and the voters of the city ratified a measure which authorized a Board of Trustees, appointed by the Superior Court of Cincinnati, to borrow money on the city's credit and to build a railroad to Chattanooga. It might well be, as the Cincinnati *Daily Times* said, that improvements could be better constructed and managed by private companies than by municipal authorities, but "extreme cases need extreme remedies," and securing the railroad was "a matter of life or death." [17]

The popular vote was taken in 1869, but a long delay in obtaining the consent of the Kentucky legislature and other difficulties prevented the actual initiation of the project until after the panic of 1873. The effect was to make borrowing difficult; the sums actually spent, which ultimately amounted to $18,000,000 for the railroad and $300,000 for the Cincinnati terminal, were obtained by the sale of irredeemable bonds at interest rates running as high as 7.3 percent. Construction of the 334-mile railroad began in 1874 and was completed, after considerable hesitation and controversy, in 1880. In the following year the road was leased to a private company which assumed the responsibility for bringing the road into good condition and for annual rentals which were to increase over a period of twenty-five years from $800,000 to $1,250,000. These sums would not until near the end of the period fully cover the service of the city's railway loans, which for several years had been absorbing more than a third of the municipal tax revenues. The bids, however, exceeded expectations; and, on the occasion of the lease, the Cincinnati *Commercial* went so far as to declare that the Trustees, "with all the millions they have handled . . . have yet to get the worst of it in a money transaction." [18]

From the time of its opening, the railway carried substantial traffic, and it soon became, as it remains, a part of the "Queen and Crescent Route" between Cincinnati and New Orleans. After an examination of freight rates and of the volume of trade with the South, the Cincinnati Freight Bureau attempted in 1892 to assess the railroad's contribution to the city in monetary terms. "There should," it said, "be entered to the credit of the Cincinnati Southern Railway an unrecorded dividend of more than ten millions of dollars." [19] Whatever the precise calculation, there can be little doubt that this municipal enterprise succeeded in achieving its developmental purpose.

In most cases local aid was given in more diffuse and much less lordly fashion. In the years 1866–73, the legislatures of twenty-nine states granted over 800 authorizations for aid by local governments to railroad projects. Reports of affirmative action by about 200 local governments in twenty-seven states for aid to 138 railroad projects reached the columns of the *American Railroad Journal*.[20] How great a total was actually appropriated, in these and other cases, and to what effect, still frequently remains obscure. But much of the character of the local aid movement, in this period of greatest activity, may be learned from the experiences of three states with unusually large programs, which have been the subjects of patient scholarly investigation.

The largest expenditures of all were made in the state of New York. Between 1861 and 1875, as Harry H. Pierce has shown,[21] some 300 municipalities contributed more than $33,000,000 in support of railroad building, mainly in the form of stock subscriptions. To this the state added almost $1,200,000 in outright gifts, after the railroad advocates had pointed out that the constitution forbade loans and subscriptions but said nothing about donations.

The largest single recipient of municipal aid was the New York and Oswego Midland, to which nearly fifty towns subscribed about $5,700,000. In this case inhabitants of communities that had been by-passed by the major routes listened willingly to the persuasive arguments and "flowery speech" of the company's promoters and particularly of its president, De Witt Littlejohn, speaker of the State Assembly, whose name was "itself a tower of strength in the Midland counties." Pierce describes the railroad as zigzagging across the state in search of the municipal bonds which were its principal source of capital. Syracuse, which refused to subscribe, was left on one side; and the Midland as finally constructed managed to cross the breadth of the state for nearly 250 miles from Cornwall on the Hudson to Oswego on Lake Ontario without passing through a single major city. Even Utica and Rome were reached only by branches, built mainly with municipal subscriptions. There were critics and opponents, but most of the voters appeared to favor "anything in the form of a railroad." "Whether it goes crooked or straight," said a contemporary letter, "they seem to have no care." [22] The road's completion, which took place in the crisis year of 1873, was hailed in the railroad press as a major achievement.[23] The company, however, went into bankruptcy within a few months of the opening; and most of the investment of the towns was wiped out. The successor company, the New York Ontario and Western, did little better. A final commentary is provided in our own day by the abandonment of the route in 1957 in spite of the efforts to maintain it made by civic bodies in the towns along the line.

The history of the Albany and Susquehanna resembles that of the Midland in promotional methods and in popular response, but not in its economic and financial outcome. James H. Ramsay, its president and the chairman of the committee on railroads of the State Senate, played a role corresponding to that of Littlejohn and collected a substantial commission on the municipal bonds secured through his own "indefatigable and persevering" efforts.[24] Town subscriptions totaled $950,000, but in this case they were supplemented by a loan of $1,000,000 from the city of Albany and a donation of $750,000 from the state. The 142-line road from Albany to Binghamton was completed in 1869. It became the object of a contest for control between the Erie and the Delaware and Hudson and then an important part of the latter's system. Most of the towns were able to sell their stock at par or better, and the road performed a substantial function as a carrier of coal from the anthracite field.

Of the many other railroads aided by municipal subscription, several were projected, like the A. & S., largely to bring coal from the Pennsylvania mines. In a few cases, the roads for which the towns bonded themselves were never built. In a few others, the municipalities ultimately secured large financial returns. Much more often, the towns lost money on their immediate investment but secured useful railroad connections. The depression of 1873, during which more than three quarters of the publicly subsidized railroad mileage of the state went into receivership,[25] brought about a revulsion against railroad aid: in 1874 the voters approved an amendment which made unconstitutional loans, subscriptions, or donations by either the state or its municipalities.

In Illinois, which had an early and abortive program of state aid followed after 1848 by a relatively moderate amount of local assistance, there was, as in New York, an extraordinarily active period of local government activity in the years after the war. The figures assembled by Kathleen Bessie Jacklin show a total of almost $18,000,000 appropriated after 1861, almost all in the form of stock subscriptions. Though Chicago did not take part in the movement and was described by a downstate paper as "fattened . . . by the railroads other counties have built," participation was widespread through most of the state. Well over fifty counties and almost 300 municipalities are recorded as authorizing railroad aid.[26] The characteristic concerns of the communities are illustrated by the conditions placed on some of the subscriptions. Squaw Grove Township, for example, provided that the route must be located within half a mile of the center of the town and that the road be built with iron of a given weight within a given time; Saline County made the condition that the railroad receiving aid must pass within half a mile of the courthouse in Raleigh and of the church in Galatia. A few of the towns used their subscriptions as inducements to secure location of railroad repair shops. The river town of Quincy, which made the largest municipal suscriptions, supported construction on the other side of the Mississippi as well as within the state of Illinois in order to establish and enhance its position on a major line of transportation to the West.

The movement for local aid received additional stimulus from a law adopted in April, 1869, offering tax advantages for counties and municipalities that bonded themselves for aid to railroads. Toward payment on these bonds the State Treasurer was for ten years to apply all of the general state tax received on *increased* assessments in the bonding localities, and the local authorities were required to apply to the same purpose the taxes they collected on the railroad property. The measure, which may be thought of as a rough attempt to appropriate to railroad building some of the "inappropriable" gains created by the process, had the effect of offering state aid without state control or supervision. A number of localities hastened to take advantage of the opportunity to further their favored improvements largely at the expense of another agency. Some $5,000,000 of new bonds were registered under the act before the end of 1870. On the other hand, citizens of areas which were not giving subsidies, led by the Chicago press, campaigned against what they considered the manifest inequity of this "tax-grab law." The entire question of railroad aid was the subject of heated controversy in the constitutional convention of 1870. In July of that year, by special vote, the citizens of the state approved a provision forbidding further local aid, though permitting the issuance of bonds under authorizations already voted. "By a vote of four to one," said the Chicago *Tribune*, "they have put to an end that kind of public robbery and local impoverishment." [27] By this time the local authorities had indeed assumed very large obligations, which bore on them still more heavily when the use of the state tax money was later held illegal; but they had in the process contributed substantially to an extraordinary burst of railroad building. The figures of construction followed those of subsidy, and the more than 3,000 miles built in the five years ending with 1872 nearly doubled the railroad mileage of the state.

In Missouri, the enthusiastic resort to local assistance during these same years was in sharp contrast with the abandonment and liquidation of its program of state aid. By 1861 most of the railroads that had received public aid were in default on their obligations, and little construction could be accomplished during the war years. The state government refused to make new appropriations for railroad aid, but it showed greater interest in the completion of the roads than in the protection of the investment it had already made. Legislation in 1864 and 1865 subordinated the state's claim to a second lien on one railroad and part of another in order to permit the companies to raise capital by the sale of first mortgage bonds. Later in the decade, after the state had taken over most of the defaulted lines, it sold these roads and its interest in the Pacific Railroad to various purchasers for a total of little more than $6,000,000. This amount was less than a fifth of what the investment had cost the state in principal and interest and was considerably lower than the capitalized value of the current earnings of the companies. The purchasers, however, accepted obligations to complete the projected lines, and in the case of the St. Louis and Iron Mountain an additional $665,000 was returned to the company to secure the building of an extension. These obligations were fulfilled by the building of 626 lines of new railroad, as against 914 previously completed by the state-aided companies, and the earlier subordination of the state's lien had aided in the completion of the Pacific Railroad to Kansas City. Yet a report made by a lobbying committee to the board of directors of the Pacific Railroad Company suggests that legislators were influenced by other considerations as well. Since millions were at stake, the committeemen had resorted to what they regarded as "the usual mode" of persuasion. Their success in obtaining for $5,000,000 release from the state's claim of nearly $11,000,000 had, they said, cost them "an amount of anxious labor, pain, mortification, and we may safely add degradation" for which they could never be adequately compensated.[28]

At the same time that the state government was ridding itself of railroad entanglements, the counties and municipalities of Missouri were enthusiastically adding to their commitments. There had been substantial local government aid to railroads during the fifties, but efforts were redoubled in the later period with subscriptions of nearly $18,000,000. This included $700,000 appropriated by St. Louis County in 1864 to aid in the completion of the Pacific Railroad. Almost all of the remainder, as Edwin L. Lopata has shown, was subscribed during the boom of 1865–72.[29] The *Missouri Democrat* reported public celebrations of favorable decisions, with serenades for the principal proponents; and the *Missouri Republican*, which spoke in 1868 of "the railroad mania," was still tracing the spread of "the railroad contagion" in 1871.[30] Nearly fifty counties and a still larger number of municipalities took part in the movement, stretching their credit by floating bonds at interest rates up to ten percent, which the railroads often could sell only at discounts of twenty-five or thirty cents on the dollar. Much of this aid was given to locally owned roads intended to provide connections with the main trunk lines. Most of them, however, eventually became parts of larger combinations, and nearly $7,000,000 of the total was subscribed to companies brought together in the Missouri Pacific. At the end of a period of active railroad building, new subscriptions ceased almost completely after 1872; and in 1875 a provision forbidding local as well as state aid was included in the Missouri constitution.

Each of these three large programs was brought to an end in 1870 or shortly afterward, and in each the door to future local aid was closed by constitutional amendment. The same is true, as has been shown, of the four Southern states of Arkansas, Alabama, Florida, and Texas.[31] In each case, except that of New York, the program of assistance slowed down or came to a halt before the panic of 1873; but, except in Illinois, the constitutional provisions were not enacted until the feeling of disillusionment was strengthened by the widespread railroad bankruptcies of the depression that followed.

In a number of other states, local aid and the revulsion against it followed a somewhat similar time pattern. In Michigan, court decisions in 1870 brought the practice to an end by holding that it was subject to the constitutional ban against state action. In Wisconsin, in the years following 1868, the frontier communities of the northern half of the state bonded themselves as enthusiastically for railroads as had those of the southern half in the years before the war. Construction in the area was also aided by federal land grants, and the northern representatives in the legislature came close to victory in an attempt to remove the prohibition against state aid.[32] Here as elsewhere, however, the period of active local aid was over by 1873; and the practice was restricted, though not prohibited, by constitutional amendment in 1874.

In California, city and county aid to the Central Pacific and

Western Pacific totaled $1,600,000, as has been shown; and subscriptions were also made to other railroads. San Francisco's aid to the Central Pacific was the subject of particular controversy. A subscription of $600,000 was approved by the voters in an 1863 referendum, in the course of which agents of the company were said to have distributed five- and twenty-dollar gold pieces in the streets. After the decision had been challenged in the courts, a compromise solution was reached which throws light on the current belief in the value of the stock. Under it the city was allowed to substitute a free gift of $400,000 for the subscription of $600,000. An act of 1870 authorized counties to grant railroad aid up to five percent of their assessed valuations; and the city and county of Los Angeles, at a time when the city's population was 6,000, obtained their connection with the Southern Pacific by making a donation in 1872 of about $600,000. In the same year the law was repealed and the period of local aid came to an end.[33]

On the other side of the continent, the New England states also shared the enthusiasm for local aid in the later sixties and early seventies. In New Hampshire and Vermont a number of the hill towns, abandoning their characteristic ante-bellum frugality, bonded themselves heavily in order to secure railroad connections. In New Hampshire, Dover, Portsmouth, and Nashua made substantial subscriptions. Local aid was given by several of the smaller Rhode Island towns and also, as has been shown, in Massachusetts. In Connecticut, town aid, given by some thirty communities, provided more than half of the increase in railroad capitalization in the decade beginning with 1868. By 1875 the railroad debt of the towns was over $5,500,000 including about $3,000,000 incurred by the cities of Hartford, Middletown, and New Haven. The railroads aided from these various sources were many of them short local lines, and few of them survived the decade of the seventies without default.[34]

The state of Maine was again exceptional for the boldness of its projects.[35] John Alfred Poor's dream of the European and

North American Railroad was finally achieved in 1871 (though without revolutionizing the routes of trans-Atlantic traffic) by the completion of a through line from Bangor to St. John, New Brunswick. To accomplish this required the piecing together of public aid from a surprising variety of sources. A sum of about $850,000 was obtained by assiduous lobbying in the three capitals of Washington, Boston, and Augusta. Long-standing claims against the federal government, based on expenses incurred in the War of 1812 and on losses of land and timber resulting from the settlement of the disputed boundary with Canada, settled by payments to Massachusetts and Maine, and the states transferred the proceeds to the company. In addition, Maine added the gift of some 700,000 acres of public land, on part of it reserving the timber rights; the city of Bangor made a loan of $1,000,000; and the Canadian authorities subsidized the road on their side of the border.

The city of Bangor also provided most of the funds for the Bangor and Piscataquis, loaning $925,000 and purchasing $200,000 of the common stock, and it took the principal responsibility for its management. Built northwards into the Maine forests, the road long remained a drain on the city's finances. Smaller ports harbored similar ambitions. During most of the sixties, Calais owned and operated a railroad to which it had made an earlier subscription. Belfast was the majority stockholder in the Belfast and Waterville. Bath joined a number of other towns in providing most of the capital for a railroad that struggled along the rocky and indented coastline. Meanwhile Portland, not content with its ante-bellum achievement with the Grand Trunk, invested heavily in two other projects for major routes to the West. One, the Portland and Ogdensburg, was to reach Lake Ontario, passing through the White Mountains by way of Crawford Notch, which "the Great Architect of the Universe" had left open, as the Portland *Press* assumed, "for a worthy purpose." The other, the Portland and Rochester (New Hampshire), had no such predestined "channel"; [36] but its projectors planned to by-pass Boston and establish connections with New York and the West. The city's investment, beginning with an 1867 referendum which allotted $700,000 to each enterprise, rose to more than $2,000,000 in the Portland and Ogdensburg, including the majority of the common stock, and to well over $1,000,000 in the Rochester. Each of the roads was completed during the depression to meet its western connections, but neither fulfilled the hopes of substantial through traffic, and both were bankrupt by 1877.

Local aid in New England thus extended into the depression years of the seventies, after it had been abandoned in many parts of the country, but it came to a virtual halt before the end of the decade. The change was marked in 1877 by Portland's sale at a heavy loss of its interest in the Portland and Rochester and by New Hampshire's constitutional amendment prohibiting local aid. Connecticut took the same step a few years later. Only in Maine was there to be any substantial later revival.

In still other cases, the timing of railroad aid programs was largely controlled by the westward extensions of the railroad network. In Minnesota the city of St. Paul appropriated $500,000 for railroad aid in 1864, and a number of other authorities gave assistance during the boom period before 1873. After the depression, the practice was revived with a general law of 1877 governing the terms of municipal aid; and during the next ten years there were also a number of measures authorizing assistance to particular railroads by counties, by Sauk Center and other towns, and by villages such as Young America. The total amount of local bond issues, from 1869 on, approached $3,000,000.[37] In Iowa the granting of local aid was interrupted by an adverse court decision in 1869 and by several shifts in policy; but the total amounts were substantial, and the practice was widespread. Municipal bodies in at least twenty counties voted local aid before 1873, and the movement spread to twenty-two additional counties before the end of the century.[38] The state of Kansas, according to its financial historian, "seemed to be the hotbed of the disease" of railroad aid. Subsidies from local governments exceeded $4,000,000 in the years 1870–78 and were more than $12,000,000 in the years 1879–90. In the second of these periods over 6,000 miles of railroad were constructed within the state. The abandonment of some of this mileage during the following decade appears to bear out the charge of over building.[39]

Railroad assistance was a controversial issue in Nebraska and in the territory of Wyoming, and Union Pacific officials were described as taking an active part in local elections in support of appropriations for branches of their own system or in opposition to those for rival roads.[40] In 1867, the Nebraska legislature authorized Omaha to aid in the construction of a railroad bridge over the Missouri River. A general law of 1869 empowered cities and counties to give assistance to works of internal improvement; the privilege was extended to the smaller municipalities in 1879. In forty-three counties the voters approved the issuance of nearly $5,000,000 of subsidy bonds.[41] In Wyoming, which obtained its east-west transportation from the main line of the Union Pacific, the territorial legislature authorized local assistance to encourage the building of railroads running north and south and northwestward into Montana.

The legislature of Dakota Territory passed a general act in 1871 permitting railroad aid by local governments, and there were also a number of special laws. The promoters of the Dakota Southern included the political leaders of the territory, and most of the actual cash investment was furnished by Yankton County with a bond issue of $200,000 and the town of Elk Point with $15,000. But Yankton County soon attempted to repudiate its bonds, carrying its appeal to the United States Supreme Court in 1880;[42] and the practice of railroad aid was not continued after North and South Dakota attained statehood. Though South Dakota adopted in 1889 a general law authorizing municipal aid, its constitution of 1891 prohibited all forms of state and local assistance.[43]

A block of states in the Mountain area and the Pacific Northwest made little response to the proponents of public aid. Colorado, which authorized a limited amount of local assistance during its territorial period, entered the Union in 1876 with a constitution forbidding both state and local aid. In Montana the record is entirely negative. The promoters of the Northern Pacific hoped that they might obtain $1,500,000 from the territory,[44] and the legislature passed laws prescribing the terms under which counties might aid certain railroads and the territory itself might aid the Northern Pacific. But none of these proposals received the required majority vote of the electorate,[45] and the constitution of 1889 prohibited railroad aid by either state or county. In Idaho, except for a measure granting tax concessions, there were no laws authorizing railroad aid.

The delegates to Oregon's constitutional convention in 1857, impressed by the failure of railroad aid programs in the Middle West, adopted provisions forbidding local aid and making state action almost impossible. In 1866 the legislature attempted to evade these restrictions by agreeing to pay interest on $1,000,000 of the bonds of the Oregon Central, and in the next decade the Central Pacific Railroad offered to give Portland a connection with the East on condition that the state and city reverse the previous policy and join in granting a massive subsidy. But the state did not pay the interest, the Central Pacific offer was refused, and the only aid actually given by the state consisted of the transfer to certain railroad companies of land granted by the federal government and the proceeds of the sale of federal lands.[46] In the state of Washington, it was the city of Seattle that offered and the Northern Pacific that rejected, a substantial subsidy on condition that Seattle be made the western terminus of the road. Two of the Puget Sound cities made attempts to build railroads by the voluntary effort of their citizens, and several counties in eastern Washington voted support to railroads toward the end of the eighties, but the total amount of public assistance was relatively small.[47]

The four "Mormon Railroads" of Utah, built in the decade following completion of the first transcontinental, were a special case. They were constructed almost entirely without government funds and indeed were hailed as notable examples of unsubsidized private enterprise. Yet in fact they represented the collective effort of a tightly-knit religious community. The leadership and much of the investment came from Brigham Young as "trustee-in-trust" for the Church of Latter Day Saints, and a large part of the labor and supervision was furnished directly by its "wards" (dioceses) and local congregations.[48]

In Nevada, a number of proposals for railroad aid were vetoed by an antisubsidy governor; but counties provided part of the funds that built two roads into the mining regions, which earned good returns until the silver began to run out. For a third road, the Nevada Central, the subsidy was made effective only by a memorable example of municipal resourcefulness. In 1875 a law was passed over the governor's veto authorizing Lander County to subscribe $200,000 on condition that the railroad should be completed to reach the county seat, Austin, not later than February 9, 1880. Construction was greatly delayed, and, by noon of the appointed day, the contractors had still two miles to go. In the afternoon, the common council of Austin met the emergency by adopting an ordinance to extend the city limits by half a mile; and the workmen, laying track on ice and snow, reached the new target ten minutes before midnight![49]

In the territory of New Mexico, several counties voted railroad aid under a general law adopted in 1872. In Arizona, a series of special laws enacted between 1879 and 1885 authorized county aid. Under one of these measures, which provided for assistance at the rate of $3,000 for a narrow gauge mile and $4,000 for one of standard gauge, Yavapai County issued $291,000 of its bonds for the Prescott and Arizona Central.[50]

Even in much older parts of the country there were a number of states in which local aid continued, or was resumed, after the depression of the early seventies. In Virginia, the Carolinas, and Mississippi, as the preceding chapter has indicated, there were many cases of local government action in the eighteen-eighties. West Virginia authorized local aid under special laws in the immediate post-bellum period; and subscriptions of $1,700,000 from five of its counties, with similar aid from the Virginia side of the line, helped make it possible for the Chesapeake and Ohio to reach the Ohio River in 1873. In the same year, when revulsion against public aid was strong in other parts of the country, West Virginia adopted a law permitting any local government authority to submit to its voters proposals for railroad assistance. Most of these later proposals, however, were defeated at the polls.[51] In neighboring Kentucky, eleven special laws were enacted in the years 1866–74 and no less than eighteen in the five years beginning in 1886. Louisville, which had made large railroad subscriptions before the Civil War, added an additional $2,500,000 in the late sixties and early seventies. Much of the legislation during these years, however, authorized aid on the part of smaller municipalities or, more often, of counties. A law of 1866, which appointed "the presiding judge and the circuit and county court clerks" of nineteen counties as "a board of internal improvement for their several counties," suggests the rural character of many of the areas concerned.[52]

Perhaps more surprising was the continuation of local aid in Indiana, most of whose neighbors had adopted constitutional prohibitions. Between 1865 and 1875 subscriptions or donations were ordered in eighty-nine municipal or county elections, and similar action was taken by forty-three townships and two cities between 1876 and 1886, with the total voted exceeding $6,000,000.[53]

In 1890 laws authorizing municipal or county aid to railroads, or more often both, still remained on the statute books of fourteen states. Since by this time the railroad network of the United States was almost completely filled in, it appears unnecessary to attempt a survey of the occasional instances of further action under these statutes.[54] There were, however, two later cases, at opposite ends of the country, which should be cited to illustrate the persistence of aid policies as long as frontier conditions continued to exist.

In Maine, the decade of the nineties marked not the end but a marked revival of railroad construction and public assistance. Some 550 miles of road were built in the northern and eastern parts of the state with substantial aid from county governments.[55] The smaller and less successful venture was that of the Washington County Railroad, which was built along the coast eastward from Mount Desert, taking over certain earlier projects, and was completed in 1899. Much of the capital was furnished by Washington County's subscription of $500,000 in 1893, and this investment was lost in a bankruptcy that preceded the road's absorption by the Maine Central. The greater project was the Bangor and Aroostook. Maine's northernmost county, which had recently been crossed by a branch of the Canadian Pacific built under Dominion subsidy, had no direct rail connection with the rest of the state. Detailed proposals for such a connection appeared in the Presque Isle *Star-Herald* in 1890 [56] and received strong local support. Legislation was obtained authorizing a county subscription and providing almost complete exemption from state taxes for twenty years. Aroostook County took $728,000 of the preferred stock (more than half the issue), and the common stock went to the construction company. By 1899 the railroad was completed to its northeastern terminus on the St. John River, and the company had relieved the financial burdens of the city of Bangor by taking over the Bangor and Piscataquis Railroad. Further construction was continued until 1912, bringing the total mileage to 630. The company became highly successful and was able to retire its preferred stock, including the county's, as early as 1904. Nor can there be doubt that it accomplished important developmental functions, opening up large areas of timberland, making accessible mill sites for paper factories on the Penobscot and other rivers, and above all making possible the exploitation of Aroostook County lands as a major potato-growing area.

No comparable claim of importance can be made for the eighteen miles of the City of Prineville Railroad, but its case appears worth citing because of the late date and the circumstances of its construction. Prineville is a community in a sparsely settled area of central Oregon. Left at one side by the railroad builders, its citizens decided to construct their own road. They were able to do this on the basis of a 1915 law which authorized cities and towns with a population of more than 1,000 to own and operate railroads, as well as water works and light and power plants, "for the benefit and use of the inhabitants thereof, and for profit." Though Prineville's population was not much above the required minimum, bond issues amounting to $385,000 were approved in a series of almost unanimous votes. Construction, delayed by the difficulty of obtaining steel during the First World War, was completed in 1918. Operation was at first unprofitable, and the city attempted in 1934 to protect its investment by imposing a special tax on those of its businessmen who gave their freight to motor truckers rather than to the railroad. There was, however, a gradual increase in traffic, particularly in lumber with a cluster of sawmills in Prineville itself; the financial position improved; and the railroad remains today, with the bond issues paid off, a going municipal concern.[57]

The City of Prineville may perhaps seem to belong rather more to the twentieth-century movement for municipally owned utilities, in spirit as well as timing, than to the nineteenth-century tradition of railroad subsidy. Yet northern Maine in the eighteen-nineties and central Oregon two decades later were both belated frontiers, and the decisions of local governments to support internal improvements under these conditions represented the continuation of a process that had accompanied the advance of settlement through the entire history of the country. The correlation between public assistance and frontier conditions, to be sure, was by no means complete. The Mountain and Pacific areas made less use of state and local aid than the areas of earlier settlement had done. On the other hand the persistence of large public programs in Massachusetts, New York, Maryland, and Indiana can hardly be regarded as a frontier phenomenon. Even in these cases, however, the principal beneficiaries of aid—such as the Tunnel Route and the Midland of New York and the Western Maryland—represented construction in the less prosperous and the less "favored" parts of the states. In general, as this study has shown, the practice of resorting to public credit lasted longer in the less developed or the less wealthy areas. An 1873 law permitting the citizens of the mountain counties of North Carolina to pay their railroad aid taxes in "bacon, wheat, beef, pork or corn," illustrates recourse to public aid as an expedient for stretching the limited means of an agricultural area remote from markets. The persistence of state aid in the South when it had been largely abandoned in the North, and the long continuance of local aid in certain Southern and Western states, are in part to be explained by their distance from the principal sources of capital.

It is of particular interest to note the continuance—or rather the persistent recurrence—of programs of local aid long after the practice of federal subsidy was abandoned in the early seventies. Whatever general views were held on the role of government intervention, the local authorities in many parts of the country found no compelling doctrinal arguments to deter them from investing public money in the railroads which they considered necessary for the development of their areas. The same traditional willingness to use whatever means seemed most convenient in the particular case extended on occasion to fully public enterprise. Though "direct action" in railroad building on the part of the federal government was clearly anathema, no such prohibition served to inhibit what might be called the Chamber of Commerce socialism of the city of Cincinnati.

To the student of railroad or general economic development, however, state and local aid after the Civil War is of somewhat less direct interest than to the student of economic ideas. Though the absolute figures of investment by these authorities are high during the years 1868–73 and though it played a considerable part in the extension and the thickening of the railroad network during that period, the ratio of public to private investment was much less than before the war. When railroad building revived after the depression of the seventies, the public share in the total of new railroad investment became almost insignificant. In the development of the greatest post-bellum railroad systems, state and local aid no longer played the indispensable part that they had taken in the construction of the Pennsylvania or the Baltimore and Ohio or other ante-bellum roads, or that federal aid was taking in the construction of the first transcontinental. Action by state or local authorities continued to be decisive in such cases as the Western Maryland, the Tunnel Route, and the Bangor and Aroostook; but much of their effort was expended on still smaller railroads, opening up new or neglected or economically unpromising areas, and often becoming branches or feeder lines of the great systems.

The relative strength and importance of the company and the subsidizing authority were very different from what they had been in earlier periods. When Gallatin prepared his plan for federal action, when New York State built the Erie Canal, when Virginia created its Board of Public Works, there was no corporation which could command anything like the resources of the government concerned. But the major railroad companies of the post-bellum period were themselves great financial powers, with resources greater than those of some of the state and territorial governments with which they dealt and certainly greater than those of most of the local governments from which they accepted or demanded aid.

The result was a change in the nature of the bargaining relationship. Certain leaders of railroads that were already well established began to attack the principle of subsidy, particularly as applied to new competing lines. Some were even doubtful of the value of subsidies for branches of their own systems.[58] Many railroad leaders, however, continued to expect and rely on public aid for part of their financing; some were prepared to go to great lengths of propaganda, persuasion, and political influence in order to secure it. The time had long since passed in the country as a whole when, as one politician said of the state of Texas, no railroad had enough money to buy "a single member of the legislature, even at a very small price," [59] and examples of corrupt influence on legislatures and on local elections can be taken from Northern states as well as from the Reconstruction South.

Bargaining between railroad promoters and local authorities often turned on the question of the location of the road and sometimes, as has been seen, on the location of its shops or other facilities. In many cases, though "nature and the God of nature"—or the principal promoters—had "imperiously fixed the general route" of a projected railroad, its line could nevertheless "be diverted by the forecast and enterprize of men, some twenty or more miles either way, without peculiar public inconvenience." [60] There was room, therefore, for competitive bidding on the part of the interested communities. A public meeting in North Liberty, Iowa, illustrated the point by adopting a resolution, "That Johnson County donate half a million dollars rather than this Rail Road should be made twenty miles east or west of us." [61] In some cases, the communities themselves took the initiative in proffering aid on condition of location. In many others competitive bidding was organized and stimulated by the railroad promoters. "The builders . . . agree . . . to make Covington a point if she will make a donation of $80,000, otherwise they will run the road in the direction of Perryville." [62] The editor of the New Albany *Ledger-Standard* urged his community to outbid Louisville in its municipal subscription in order to obtain the location of a railroad's shops and machine works. It was, he said, "business-like and perfectly proper" for the company to claim such aid, "in view of the great advantages they will bring to the city that gets them." [63]

There can be no doubt that such considerations were often decisive in determining the detail of railroad location. Two cases from Iowa, one in which the town of Vinton secured a branch line by voting a larger subsidy than the neighboring community of Shellsburg and the other in which Council Bluffs gained an important advantage over Sioux City,[64] could be matched by many from other states. Sometimes agreement was not reached. A Northern Pacific official described Seattle as "bidding up pretty well" but not enough to overcome Tacoma's advantage in distance.[65] In the case of Syracuse and the New York Midland, it was the company and not the municipality which lost when the refusal of a subsidy led the directors to leave the city "widely to the left of the line." [66] For many a prairie settlement, on the other hand, the difference of a few miles in the location of a railroad was decisive for its entire future. If the line passed the other way, "business men removed from the place and took their wares to the new city." But if it came, "a hitherto quiet and plodding village became full of the bustle of business and enterprise." [67]

Both parties to the transactions showed their recognition of these relationships. One railroad executive cautioned another that local aid must be obtained before the chosen route was announced. "Nothing can be realized from any locality after they are sure of the road." [68] The communities on their side made their aid conditional on increasingly detailed specifications. The track must run through the center of the town, or within half a mile of a certain church, or to a particular street, or even through the courthouse square; and it must be completed "in and through" the township or, as in the case of the Nevada Central, to a particular city by a given date. Provisions like these were designed to make sure that the communities' contributions would not be handed over unless the road was actually built, "and cars running," by the desired route. Such considerations were much more decisive than any hope of dividends from railroad stock, on which so many communities had been disappointed in the past, or than any desire to influence company policy by representation on its board of directors, which was in any case hardly practicable where the aid was pieced together from the contributions of dozens of villages and townships. Under these conditions, the acquisition of railroad stock often seemed unimportant to the local government authorities; and in some states the conditional donation replaced subscription as the most common form of local aid. "Does the company have to give stock," asked a railroad president, "or is it a square donation?" [69] If the second alternative seemed more likely to induce the company to build the desired line, many a community was willing to give the answer of outright subsidy.

In earlier decades, ambitious states and cities had initiated major projects or acted as equal partners in mixed enterprise. By contrast many of the local governments in the post-bellum period were very junior partners indeed in the enterprises of great corporations. Baltimore continued and Cincinnati revived something of the tradition of bold municipal enterprise to obtain "those results most desirable to a commercial city—increase of trade and cheapening of transportaton." [70] Much more often, however, local aid had merely the aim of winning for the community a place—or a particularly favored place—in a railroad network of which the main outlines were determined by forces entirely outside the control of the local authorities. To say this is not to deny the importance of the decision to the community in question. For many a "Center Township" or "Smith's Four Corners," railroad aid was no doubt more literally "a matter of life and death" than the building of a Southern connection was to the city of Cincinnati. Yet the township donation hardly represented government planning of development. In these later decades of internal improvements, the scope of public decision was much narrower than it had been when the state of New York built the Erie Canal and the emulation of states and cities played so large a part in breaching the Appalachian Barrier.

CONCLUSION

. . . this remarkable movement toward State enterprise here in America, where of all places in the world we should least expect to find it.

Guy Stevens Callender, in a 1902 article

8

PUBLIC PROMOTION AND PRIVATE ENTERPRISE

The preceding chapters have presented an account of the part played by public agencies in the promotion of canals and railroads. They describe a movement that continued, or persisted in recurring, throughout the nineteenth century and that engaged the effort of an extraordinary diversity of governmental agencies. No single explanation can account for all the variations of public action; they offer a rich field for the analysis of the subjective as well as the objective factors determining political and economic decisions.[1] In general, the inception and abandonment of improvement programs can be related to the stage of development reached by the particular area at the particular time, to the obstacles or opportunities presented by its geographic features, and to the abundance or scarcity of private capital available for investment. The timing of the decisions was often dramatically linked to the ups and downs of the business cycle. In some measure also, the variations reflected differences in the fortunes of opposed political parties. In national politics, John Quincy Adams formulated the most "enlarged" philosophy of internal improvements; the Jacksonians put an end to the Era of National Projects; and Southern Democrats provided the most implacable opposition to the revival of national subsidy in the eighteen-fifties. Yet it must be remembered that the National Road was a Jeffersonian project and that the two most comprehensive federal programs of improvement were put forward by Albert Gallatin as President Jefferson's Secretary of the Treasury and by John C. Calhoun as President Monroe's Secretary of War.

In state politics, the suggestion that Whigs favored and the Democrats opposed internal improvements should be viewed with equal caution. Revulsion against public construction in New York became a matter of party doctrine for one wing of the Democrats, and similar instances may be found in other states. Yet in still others the position was reversed, with Michigan Whigs campaigning successfully against a Democratic program of improvements and Georgia Whigs providing the chief opposition to expenditures on the construction of the state railroad. Even in Southern Reconstruction, railroad subsidy was not exclusively a doctrine of the Radical Republicans nor was revulsion against it confined entirely to the Redeemer Democrats. In most cases, programs of state action had at their inception wide popular support regardless of party; the differences that developed were more often traceable to conflicts of local and sectional interests than to differences in party doctrine. Such interests were expressed still more directly in the aid given by local authorities, but in state and even in national politics internal improvements often appeared—in the words historically applied to the protective tariff—to be largely "a local issue."

Yet the public promotion of internal improvements was a nationwide phenomenon, whatever its diversities, and it played a role in the development of every part of the Union. It remains to ask what effect these measures had on the growth of the economy and what light their history may throw on the nature of American development.

A first essential is to obtain some indication of the magnitude of public aid and investment. Recent studies have greatly increased the amount and precision of the available data. Though much remains to be done in the detailed examination of state and local programs, and though systematic allowance has usually not been made for the frequent discounts and the less frequent premiums at which state and local government bonds were turned into cash for construction,[2] it now appears possible to prepare serviceable though still highly approximate estimates of the totals of direct government expenditures on internal improvements. It must, however, be recognized that much of the aid was given in forms other than cash or credit. The principal form in which support to transportation development was extended by the national government and by the state of Texas was that of grants of the public lands, and any attempt to convert acres into dollars confronts the basic difficulty that the value of the land in question depended so largely on the completion of the enterprise for which it was granted. It would be even more difficult to make accurate calculations of the value of the many other contributions that were made in kind, such as buildings or town lots furnished by local communities, or for the profitability of banking or other privileges extended to certain of the corporations, or for the benefits so frequently conferred by tax exemption.

Even where the governmental commitment consisted of a specified amount in dollars, there were wide differences in the way in which it was applied. Money might be spent directly on a public work, or aid might be granted to a corporation under various terms and conditions—subscription to stock, loan, guarantee of interest or principal or both of the company's obligations, or simply an outright donation. From the point of view of the finances of the governmental unit, any one of these operations—except a guarantee that never had to be honored—involved either current expenditure or the incurring of an obligation that had ultimately to be met (or repudiated); but its future position would obviously depend on what it obtained, or failed to obtain, in earnings or dividends or interest or loan repayment and on the capital value of its investment. From the point of view of the finances of the corporation, its future position would similarly depend on whether or not it was obliged to make payments under these various headings and on whether or not it was required, as in the case of the federal land grants, to render special services to the government. Yet there is one point of view from which these differences may be regarded as largely irrelevant. In the field of transportation, American governments almost never contributed to the resources of enterprises which were not thought of as "needing" their support. All these diverse items of governmental commitment, therefore, contributed to the amassing of capital at the time when it was needed for construction, and they may be taken together for the purpose of assessing the amount of public aid to transportation development.

In the period before the Civil War, the state governments committed some $300,000,000 in cash or credit to internal improvements, and local governments more than $125,000,000. Well over half of this amount was spent, as Chapter II has shown, in the seven states directly confronting the Appalachian Barrier. By contrast, the federal government spent on internal improvements a sum less than one percent of this total during the Era of National Projects which ended with the Maysville Road veto. Even adding the costs of completing the National Road after 1830, the direct financial expenditure of the central government on early internal improvements was only about $7,000,000, though account should also be taken of its contributions to the development of transportation through the technical assistance rendered by the Army Engineers and through improvements not intended to produce revenue, such as military roads and river and harbor improvements. On the other hand, the federal government made much larger sums available to the states for their own internal improvements programs, by payments of a part of the proceeds of public land sold within their borders and through distribution of the federal surplus revenue, which a number of states applied to improvement purposes. There were also substantial gifts of land to the states to be used at their discretion for the same purpose. The practice of giving federal land to the states for the construction of specific improvements, which began with early grants for turnpikes and the Illinois and Michigan and the Wabash canals, was revived on a larger scale in 1850; approximately 22,500,000 acres were distributed under the legislation of this decade.

In the period after 1860, there is a marked change in the relative contributions of the different levels of government. The Era of National Subsidy, which was initiated with the Illinois Central act of 1850, reached its peak in the few years following the end of the Civil War. The companies that constructed the first transcontinental railroad received loans of nearly $65,000,000—the one major case of direct financial aid from the federal government. In addition, land grants were authorized during the years 1861–72 to a number of railroad companies, from which they ultimately obtained well over 100,000,000 acres. On the other hand, state governments, many of which were now bound by constitutional restrictions against aid to internal improvements, did much less than before; their total contribution in this period was less than that of the national government. Texas matched a part of the federal land grants by adding some 27,000,000 acres to the more than 5,000,000 it had granted before the war. Financial aid by the state governments, which was for the most part confined to Massachusetts and to nine states of the Reconstruction South, appears to have approached a nominal total of $95,000,000; a substantial deduction, however, should be made for discounts on the sale of state securities and for adjustments which substituted smaller amounts of state bonds for guarantees of company obligations. By contrast local government aid during this period attained very large proportions, though the available figures are far from complete and are subject to deductions for discounts on the sale of municipal and county bonds. In the three states of New York, Illinois, and Missouri, commitments for local aid reached a total of almost $70,000,000 before their programs, like those of many other states and that of the federal government, ended in the early eighteen-seventies. Somewhat later, Cincinnati spent more than $18,000,000 in building its municipal railroad. For the entire period 1861–90, local government commitments for railroad aid may well have exceeded $175,000,000, and in a number of cases assistance continued to be given after this date.

These figures become more meaningful if they are compared with estimates of the magnitude of total investment in transportation. In the case of the canals, it is clear that public authorities provided the greater part of the initiative and of the resources. Total expenditure for canal construction from 1815 to 1860 has been estimated as about $195,000,000. Of this amount 62 percent represents public works, including the major programs of New York, Pennsylvania, and Ohio; and allowance for government participation in mixed enterprises like the Chesapeake and Ohio and Louisville and Portland Canals would bring the total of public investment to about 70 percent of the whole.[3]

For the larger magnitude of investment in railroads, the bases for comparison are somewhat less firm. In this field, government participation usually took the form of stock subscription or loan rather than direct construction. For eleven states of the South, it has been estimated that public agencies contributed well over 55 percent of the cost of all railroad construction before 1861 and at least 75 percent of the amount made available in cash.[4] Of the four trunk-line railroads that reached Lake Erie or the Ohio River by 1855, public sources had provided about half of the funds for the Pennsylvania, the Baltimore and Ohio, and the Erie but only a small fraction of the capital of the New York Central. For the nation as a whole, the proportion of public to private investment in railroads before the Civil War was considerably less than in the South. The total of public commitments in cash or credit amounted to somewhat more than $280,000,000 in addition to the value of the federal land grants. The capitalization of the railroads in operation in 1861, according to the figure commonly used, was $1,145,000,000. Though this is far from a perfect measure of the cost of construction,[5] it would appear that up to this time governmental agencies had provided at least 25 percent and more probably some 30 percent of the entire investment in railroad building.

After 1861, the relation of public to total investment is a quite different one. The greatest of the land grants were made during this period, and public financial commitments to the railroads after the war somewhat exceeded the total of the ante-bellum period, at least in face value.[6] The federal government loaned half the money with which the first transcontinental road was built, and its enormous grants of land along with those of the state of Texas formed a substantial part of the assets of a number of Western and Southern railroads. Yet for the nation as a whole, these contributions, substantial as they were, must be viewed in relation to a greatly increased volume of private investment. The direct financial contributions of government agencies to railroad construction between 1861 and 1890 were somewhat less than $350,000,000. Of this the entire federal outlay and most of the state and local commitments were made in the few years before 1873. By this time, the total reported capitalization of the railroads was some $2,600,000,000 greater than in 1861.[7] It is therefore clear that the financial contributions of public agencies, even in these years of active aid programs, played a much smaller part than before the war in the increase of railroad construction. After 1873, there was a marked change on each side of the comparison. In the wave of new investment that brought the total of railroad capitalization in 1890 to more than $10,000,000,000, public agencies contributed no more than a negligible fraction.

As these comparisons indicate, the proportion of government to total investment was greater in canal construction than in the building of the railroads which made so many of the canals obsolete. A larger total of public funds was devoted to railroads than to canals; but in this case, the proportion of government to total investment was greater in the more experimental period than when the railroad industry was well established and greater in the less developed parts of the country than in the well settled areas. Under these circumstances, it is not surprising to find that the direct effect on government finances was on the whole unfavorable. If a consolidated set of accounts could be presented for the several thousand government agencies that made appropriations for internal improvements, their receipts from net earnings, interest, and dividends would represent much less than a commercial return on the sums expended, and the balance on capital account would be heavily negative.

Less often recalled, perhaps, are the cases in which the improvements were successful even by the test of direct financial returns to the investing governments. New York State's venture in building the original Erie Canal proved a financial success within an unexpectedly brief period, though it was followed by losses on the later feeder canals. Philadelphia received good dividends in cash and stock on its investment in the Pennsylvania Railroad and sold its last shares at par in 1880. Baltimore had a similar experience with its B. & O. holdings, though it failed to take the opportunity to sell them at a substantial profit. Cincinnati's municipally owned railroad yielded a reasonable return. Among smaller local government investments, Albany lost no money either on loans or guarantees, and Aroostook County's investment in the eighteen-nineties was a profitable one. Nearly half of all the railway aid granted in the ante-bellum period went to Southern railways, and the historian of the movement declared that "there was every indication" toward the end of the fifties that the majority of them "would be financially successful and return the various public investments without loss." [8] The war's dislocations intervened before these indications could be tested, but Georgia's state railroad had already demonstrated the ability to earn a satisfactory return on its investment and was to do so again after the war, except for the year in which it was deliberately plundered by its own management. As for the federal government, it finally secured repayment in full and with interest of all but a small fraction of the amounts that it loaned for the building of the first transcontinental.

Much more numerous were the cases of financial loss or small return. Of the federal government's early canal subscriptions, only that to the Louisville and Portland Canal paid steady dividends. Pennsylvania sustained a heavy financial loss on its public works, Maryland on its investment in the Chesapeake and Ohio Canal, and Ohio on its later canals. In Indiana the extensions of the Wabash Canal were unprofitable, and both Indiana and Illinois lost virtually their entire investments in the programs of railroad building that were stillborn in the crisis of 1837. In Virginia, where the James River and Kanawha Canal "operated as a moth" on the Fund for Internal Improvements, interest charges from 1815 to 1861 exceeded the state's returns from internal improvements by $22,500,000.[9] South Carolina's struggle with the Blue Ridge was a costly operation, and Massachusetts won its final victory over the Hoosac Mountain only by taking a heavy financial loss. The state programs of the Reconstruction South, though not as disastrous as they have been sometimes regarded, were certainly much less successful than those of the ante-bellum period. Local aid programs, particularly those launched in the boom years from 1868 to 1872, typically showed financial losses in the North and West as well as in the South. "From a strictly investment point of view," as has been reported, "New York communities suffered heavily by aiding in the construction of railroads. Only 52 of the 297 municipalities that bought stock in a railroad community disposed of their securities at par or better; 162 held stock with no market value." [10]

In a number of cases, these losses brought state and local governments to the point of insolvency, and they furnish part of the explanation for a large number of cases of repudiation of governmental obligations. Pennsylvania, Maryland, Indiana, and Illinois were all in default on their improvement obligations during part of the ante-bellum period. Michigan repudiated part of its

$5,000,000 loan; and Minnesota at first refused to honor its railroad bonds of 1860 and later agreed to a compromise settlement. Eight Southern states defaulted on bonds of the Reconstruction period, and in every case the final settlement was for less than the face value of the obligations.[11] In the case of local governments, defaults and litigation occurred in no less than twenty-five states [12] and were common in each of the three largest programs. Fifty-six New York municipalities repudiated bonds carrying a face value of over $4,000,000. Of this sum, $800,000 was invalidated by the courts, and for much of the remainder the bondholders accepted less than full payment.[13] In Illinois, a still larger amount of local government obligations (about $2,000,000) was voided after extensive litigation in both state and federal courts.[14] In Missouri, a large part of the local railroad debt was scaled down, often after protracted controversy; and the officials of certain counties and municipalities found a variety of expedients for deferring payment even when the courts had made judgments against them.[15]

It would, however, be a mistake to attribute the repudiations entirely to the losses incurred on improvements or to the financial plight of the governments involved. The part of Michigan's debt which it refused to honor was that on which the bankers had never forwarded the proceeds, and those who opposed payment on Minnesota's railroad obligations called them "the swindling bonds of 1860." Fraud had tainted the railroad loans of the Reconstruction South, and the repudiating Redeemers felt little moral obligation to honor the commitments made by their hated predecessors. Repudiation by local governments was also frequently based on the ground that the aid had been fraudulently obtained or that the company had not fully met the conditions of the loan or subscription. It is not difficult to understand the strength of the sentiment for repudiation in the minority of cases in which the promised railroad was never built and in which the taxpayers asked the question, "Here's the bond, but where's the road?" [16] Yet these were by no means the only issues that influenced the actions of the local governments or the decisions of the courts. On the one hand, there were cases in which the validity of bonds held by innocent purchasers was upheld even where there had been manifest fraud in their issuance; and on the other, there were cases in which there was no valid claim of fraud or nonperformance and the bonds were invalidated because technical flaws, sometimes petty ones, were discovered in the procedures under which they were issued. Nor was repudiation necessarily an indication of financial distress on the part of the government agency involved. If there were communities, especially on the frontier, that would have been brought to ruin if their obligations had not been scaled down, others managed to escape payment where the ability to pay was not in doubt.[17]

A realization of the financial losses on improvements led in many cases to the abandonment of programs and to the adoption of constitutional restraints or prohibitions. The case for such measures was particularly strong where the losses had led the state or local government to what a delegate to one constitutional convention called "the brink of repudiation and lasting disgrace." [18] Frequently, also, the sentiment of revulsion was the more intense because the advocates of internal improvements had held out extravagant promises of fiscal as well as general economic benefits. At the outset of the Maryland program, it was said, every man dreamed that he was about to reach a new El Dorado free of debts and taxes; an early leader of the movement for internal improvements in Tennessee declared that they would yield

> a clear revenue sufficient to fill her treasury and support her civil list, as well as to provide extensively for the education of her youth; and all these without taxes on her people.[19]

When the returns from the canals or railroads, far from meeting expectations like these, failed to cover the interest on the bonds issued to pay for improvements, the governments concerned had to choose between the unpleasant alternatives, "repudiation or taxation." The latter solution appeared particularly difficult since few states had established the practice or tradition of collecting substantial general levies. New York had prudently set aside for its canal fund the proceeds of certain special taxes, but through most of the ante-bellum period it levied no direct state taxes for any purpose on either property or income. Even in 1840, as staunch a canal advocate as Governor Seward could still declare that "taxation for purposes of internal improvement deservedly finds no advocate among the people." [20] Pennsylvania, with less successful public works, added in 1844 three mills to an existing general property tax of one mill to the dollar, as well as taxes on certain incomes, in order to extricate itself from default on its improvement bonds. Illinois adopted a mill tax to help pay off its obligations after the collapse of its railroad program and a somewhat larger tax to permit resumption of work on the Illinois and Michigan canal. The state of Ohio, which levied a canal tax from the beginning of its program, and the city of Baltimore, which collected an internal improvements tax in every year from 1839 to 1878, were exceptions to the general rule.[21] Though the actual burdens in most cases do not appear large in relation to the total wealth of the communities concerned, the reaction against improvement measures was the greater because direct state taxation for any purpose was still regarded as unusual and because advocates of the programs had so often and so vehemently denied the charge that internal improvement was only "another name for eternal taxation." [22]

Yet contemporary judgment frequently did not, and the student of economic development certainly can not, accept as conclusive the test of direct financial return to the government treasury. Expenditures on improvements were often maintained and increased even after the record showed meager returns on earlier projects. The well-publicized financial losses of some of the older states did not deter most of the areas of new development from following the path of public aid. Even after the state of Virginia had sold its improvement stocks at a heavy loss, a Senate committee on the public debt recorded its opinion that the money spent on Virginia's railroads had been more than justified by their contribution to the aggregate wealth of the community.[23] In campaigns for particular improvements, their advocates sometimes obscured the issue by arguing at the same time that public expenditure was needed because private enterprise could not afford the risk and that public expenditure would yield handsome returns to the treasury. Yet in other cases the likelihood of financial loss was clearly faced. When Virginia decided to build on public account the part of the Virginia Central line that required a tunnel through the Blue Ridge, the legislators did not expect to obtain a profit for the state treasury. They made the investment precisely because they considered it too *un*profitable for the company itself to undertake. In the case of the Western and Atlantic, ultimate financial success should not obscure the fact that this part of Georgia's railroad network was built by the state because it was the one in which the cost of construction was the highest and returns most likely to be deferred. In the most important case of federal subsidy, several members of Congress who voted for the Pacific Railroad Act of 1864 made clear their belief that the government's loan would never be repaid. Where aid took the form of outright donation, as in so many of the later cases of local assistance, all thought of direct fiscal return was abandoned at the outset.

On what grounds, then, was economic justification found for these expenditures? In many cases, a part of the answer was the promise of indirect fiscal advantages. Since completion of the improvement would increase the value of property within the state or locality, the added revenue from this higher tax base would outweigh any loss on the immediate investment.[24] The more important economic argument for public expenditure ran in terms of the general benefits to the community at large that could not be fully collected either by the enterprise itself in transportation charges or by the government in tax revenues. The advocates of improvements stressed the wide diffusion of these advantages and sometimes argued explicitly that governments, and also private subscribers, would profit greatly even if they never received a penny of dividends on their investments. Enumerations of benefits and lists of prospective beneficiaries varied according to the nature of the project and the constituency to which the appeal was directed. The Maryland program of improvements would "return a rich bounty . . . by enlarging trade, creating new demand for labor, raising greatly the value of land in all directions, rewarding industry by large profits." The Chesapeake and Ohio Canal would benefit agriculture "by cheapening the carriage of its heavy products." The Erie Railroad would bring advantages to "every owner of real estate—every merchant, manufacturer, mechanic and carman." The *Southern Banner* of Athens, Georgia offered a similar list but noted particularly that the proposed railroad would save the planter from "the necessity of spending so large a portion of his hard earnings in getting his produce to market." [25] Reference to the prospect of increased land values was particularly widespread. This no doubt was partly propaganda for dispelling opposition from the taxpayers, and the *American Railroad Journal* used it as its principal argument for the equity of local aid,[26] but it may also be thought of as reflecting the prominence of the realtor-promoter in frontier development. Even more general were the promises of increases in trade, which bear witness to the leadership of merchant groups in promoting ambitious projects of improvement, which were to pay their greatest dividends "in those results most desirable in a commercial city—increase of trade and cheapening of transportation" and therefore "in advantage and net profit on the business that will be thus commanded." [27]

A rule for measuring these benefits according to the principles of political economy was offered to an improvements convention in 1836: Add to the increase in land values "the time saved by man and beast" and the reduced cost of moving freight; add also "a reasonable sum for the agreeableness of manner of transacting business" with the newer modes of travel; "then say if the interest on the capital sum these advantages are worth, exceed the interest on the capital required for the completion of the work."[28] The professional economist would make a greater effort to avoid double counting and to state the various items of the formula in separable and comparable terms, but his examination of costs and benefits would take account of substantially the same points. The case for government investment or subsidy, on strictly economic grounds, would therefore depend on judgment as to the extent by which the sum of "all that diversified good" would exceed the total costs and as to the amount of this gain that would not be appropriable by the enterprise itself.[29]

Throughout the history of internal improvements, projects offering clear prospect of early financial returns were almost universally left by common consent to private enterprise. Government action or support was demanded when private investors were unwilling to bear the entire risk or, in early decades, when the amount of capital required for a single project was beyond the means or the borrowing power of any private corporation in the United States. It is not surprising, therefore, that the improvements receiving public support showed less favorable financial results than the smaller number of purely private improvements organized for the exploitation of the more obvious opportunities. The purpose of public promotion was developmental, and an assessment of its impact must take account of these diffused gains and their cumulative effect on the growth of the economy.

These developmental gains are for the most part not susceptible of direct measurement. The building of the Erie Canal brought about a spectacular reduction in transport costs and opened up for development an area that can be somewhat closely defined. For more than fifteen years the Canal provided almost the only outlet, and for a longer period the principal outlet, for the products of northern Ohio and Indiana and the new state of Michigan. To its influence, therefore, major responsibility could be assigned for the extraordinarily rapid growth of wealth and population in that region as well as for the great increases in industry and population along the route of the canal itself. Yet even in this case, such effects as its contribution to the rise of New York City and the stimulus it gave to the growth of manufacturing by providing access to a western market, though unquestionably important, cannot readily be disentangled from a host of other influences.[30]

Few improvements wrought as definite a transformation as the Erie Canal, and in most cases their specific consequences would be even more difficult to isolate. Yet all students agree in celebrating the dynamic effect of the railroads on American development, and historians have described the characteristic process by which lines built boldly ahead of traffic, sometimes at the sacrifice of the investment of the original subscribers, came to prove their usefulness as settlement followed provision of the means of transport.[31] What needs emphasis here is the extent to which this boldness of policy rested on governmental investment and support. It is difficult to imagine what the nation's transportation system would have been on the eve of the Civil War if there had been no public subsidy. Up to that time, no canal or railroad of any great length had been built entirely without government funds, although the New York Central came close to furnishing a significant exception and several profitable shorter roads in the East had been built by purely private means. Since half the capital which carried the Pennsylvania and the Baltimore and Ohio Railroads to the Ohio River in the mid-fifties had come from public sources, it is hard to believe that they would have reached this point before the war—much less established through connections to Chicago and East St. Louis—if they had had to rely entirely on private investment. In the South, the railroad network owed still more to public initiative and capital. After the war, as has been shown, government effort played a smaller part proportionally in railroad construction; but there can be no doubt that the Union Pacific, which was built well before private interests would have made the attempt unaided, gave a notable stimulus to the development of the Far West.

These were major contributions to American development, and the achievements of public investment might well have been still greater if federal resources had been used in the early decades to carry out a national plan of internal improvements or if certain of the states had provided a sounder fiscal basis for their more promising projects. Yet few critics since John Quincy Adams have suggested that too little was spent on public promotion,[32] and many have reached the conclusion that too much was spent or that expenditure was badly directed. No developmental advantage can be claimed for the public investment in improvements that were never built at all or that were abandoned, like the tunnel in Stump House Mountain, before they reached their essential goals. A particularly heavy discount must be applied to the contribution made by the public promotion of canals, since many of them were soon rendered obsolete by railroad competition. Though neither private nor public promoters could be expected, in this or any period, to anticipate the full consequences of so great a technological innovation, there were certain canal projects on which heavy expenditures continued after the rival improvement had shown clear signs of its competitive superiority.[33] Nor can it be doubted that public aid sometimes led to the building of railroads that were almost completely useless by developmental as well as by purely financial tests.

Some duplication of effort was indeed unavoidable in programs carried out by such a diversity of governments. Gallatin's Report proposed that improvements be selected with relation to the great geographic features of the country by a Congress that was to be "superior to every local consideration." Yet most of his specific proposals failed of adoption in Congresses that were torn by conflicts of state and regional and local interests, and similar considerations played an equally conspicuous part in the debates and decisions on subsidy to the Pacific Railroad. Once Gallatin's hope—or dream—of a coherent national system of internal improvements was abandoned, no comparable degree of objectivity in the choice of projects could be expected until public aid was no longer necessary and promoters could address their appeals solely to investors who wanted to make money and did not care where that money was to be made.[34]

So long as the promotion of internal improvements was mainly a function of individual states, neither North nor South Carolina was likely to be deterred from making an assault on the eastern wall of the Blue Ridge by the consideration that the Appalachian Barrier could be more easily surmounted in Virginia and in Georgia. Similar problems arose within the several states. Rivalry between Troy and Albany was no less intense than between Baltimore and Philadelphia. If Portland obtained a railroad to the St. Lawrence, Wiscasset and Belfast were bound to fail. In this competition, towns and counties by-passed by the early improvements exerted their political influence to secure state funds for their particular projects or else committed their own credit to costly programs of local aid. Occasionally a particular combination of geographic and political conditions permitted a state to concentrate its efforts on a major project of unusual strategic importance. Ante-bellum Georgia was able to confine state expenditure almost entirely to a single railroad, so located that it could serve impartially the rest of the state's transportation network. The Erie Canal performed a somewhat similar function for New York and so did the Western Railroad for Massachusetts. On the other hand, the aid given subsequently by these same states to two less advantageously located projects, the Erie Railroad and the Hoosac Tunnel, won support largely on the argument that justice required expenditure for the benefit of areas that had not yet shared in the state's largesse. Other states adopted "omnibus" programs from the outset in order to satisfy a wide range of interests, and still others attempted to avoid the necessity of centralized decision by offering aid on equal terms to improvements in any part of the state.

The pressure of these competing interests promoted a dispersion of effort among undertakings of uneven economic worth, and similar rivalries of states and localities made it difficult to transform the many separate railroad lines into a national transportation network. The Portland promoters of the Atlantic and St. Lawrence, for example, deliberately chose a gauge different from that of Boston roads to prevent diversion of traffic to the rival port; and they took the added precaution of securing from the Maine legislature a charter provision that allowed connections with their road only "on the easterly side." In the "War of the Gauges" that broke out in Erie, Pennsylvania, the making of a physical connection between two railroads was opposed because the municipality was anxious not to lose the employment created by local transshipment and because Philadelphia and the Pennsylvania Railroad wished to keep traffic away from the roads leading to New York City.[35]

Another shortcoming was the failure to develop a workable theory of the limits of public promotion. From Gallatin's day on, the advocates of internal improvements were able to present general arguments for government action that are fully understandable to modern students of development. There, they said, potential improvements that would yield substantial net benefits to the economy as a whole that would not be undertaken by unaided private investors because the gains would come too slowly or because a large part of the gain could not be appropriated in revenue to the enterprise itself. Under these circumstances the case for public expenditure is a strong one. But for how much expenditure? If the prospective profit of the canal or railroad is not to be the conclusive test, what measure is to be substituted? How are the diffused and inappropriable gains to be estimated? And how much of a direct financial loss—in general or in the specific case—should the public authority be prepared to sustain in order to secure these advantages for the community at large?

These questions received no clear answers, but many advocates of internal improvements believed that the selection of desirable projects was made easier by resort to the method of mixed enterprise. Its advantage, they claimed, was that it enlisted private judgment and "the keenness of self-interest." Willingness of investors to put up part of the money was taken as an indication that there was at least some prospect of financial return. Indeed, the private investment in mixed enterprise might be rationalized as corresponding very roughly to the expectation of appropriable gains and the public investment to the expectation of inappropriable benefits. Reliance on such private judgment was one reason for the attempts that were commonly made to maintain some fixed or reasonable relationship between public and private contributions. The original Virginia system provided that the state should make a standard subscription of two fifths of the stock of the enterprise, leaving the remainder to individuals and local governments. More common were provisions that government aid should not become effective until a given amount of private subscription had been made or paid in or until a certain amount of construction had been accomplished by private means. Yet these conditions often proved difficult or impossible to enforce. When local enterprise failed to respond sufficiently to the stimulus of the state system, Virginia raised its standard subscription to three fifths and in practice sometimes exceeded that amount. In a number of other programs, the requirements of bona fide subscription and actual construction were often flagrantly violated. Clearly the prudence of the careful investor could not be relied on as a corrective of governmental judgment where almost no real private capital was invested, as in a number of struggling railroads in the South and West, or where venture capital bore as small a proportion to the governmental contribution as it did in the Union and Central Pacific. In the extreme cases, mixed enterprise came close to representing simply the private control of public investment.

Other defects in public promotion stemmed from various imperfections of the agencies of government. Where the method chosen was that of public works, selection of personnel was often a matter of political patronage and officials were sometimes venal. Where the method was that of subsidy or mixed enterprise, occasions for corruption were no less obvious, at least after the improvement corporations had graduated from their first innocence as infant enterprises. Newspapers in various parts of the country described, with little variation in metaphor, the descent of "cormorants" or "vultures" to corrupt members of state legislatures. The Credit Mobilier brought the use of financial inducements to the Congress itself, and bribery was occasionally reported in local improvement elections. Only the connivance of public officials, moreover, can explain some of the more flagrant cases in which companies received aid without fulfilling the conditions under which it was granted.

Even if there had been no problem of corruption, the planning and administration of internal improvements would still have placed a heavy burden on what would now be regarded as "underdeveloped" machinery of government.[36] Yet in the early days of the movement, the alternative agencies of business enterprise were even less fully developed. The federal government with the Corps of Army Engineers and Virginia with Charles Crozet as State Engineer had at their disposal men of greater technical training than the infant private corporations of the day were able to employ. There were, moreover, notable cases of technological and administrative innovation on the part of government agencies. The most important advances in the technique of canal construction were made by the self-taught engineers who built the Erie Canal for the state of New York. A number of governors' messages followed Gallatin's Plan in proposing well-balanced programs of improvement. The various boards of public works or internal improvements represented an attempt to provide agencies for planning and administration. Though many of them proved ineffectual and were soon abandoned, the Virginia Board made a pioneering contribution by the clear presentation of its own annual accounts and by requiring orderly public accounting from the canal and railroad companies in the state. The Canal Commissioners of the State of New York, the standing committees on internal improvements of the city councils of Baltimore and Philadelphia and of certain state legislatures, the Board of Trustees that supervised the building of Cincinnati's railroad and a number of other agencies would repay study as specialized organs of government developed to meet the need and challenge of improvements programs.

Yet it can hardly be denied that in general the governments of the day lacked much of what would now be regarded as the essential means for the effective conduct of such a program. Their financial support was often precarious, and projects had sometimes to be abandoned halfway—not because second thought had shown that they would not promote development but for the simpler reason that the state or municipality could no longer find a market for its bonds. The planning and operation of improvements, or the effective protection of the public interest in the conduct of mixed enterprise, were responsibilities that put a heavy and sometimes unbearable strain on the capacities of state administrations that operated with limited budgets, with little expert personnel, and without civil service traditions. When the function of public assistance was more and more left to the local governments of the smaller communities, the deficiencies of the governmental machinery became still more conspicuous. State legislatures, to be sure, attempted to restrain and protect these local bodies by limiting the amount of aid (for example, to a given percentage of the tax base) and by specifying the conditions under which the consent of the voters was to be obtained.[37] Such provisions might serve to reduce the likelihood of municipal bankruptcies and defaults, but they could neither assure to the communities a voice in the councils of the railroad which received their aid nor turn the village or the township into an effective agency for the planning of economic development.

In all these characteristics, the movement for internal improvements reflected certain broad tendencies of American development in the nineteenth century. The most general impressions left by the record of public promotion are, first, the strength and persistence of the desire to promote economic growth and, second, the pragmatic willingness to make use of whatever means seemed likely to be effective in the particular case. The variety of expedients was extraordinary. Public action was taken at every level of government, from the nation to the township, and sometimes by combinations of different authorities. Still more varied were the relationships between public and private participation. The improvements were constructed sometimes as purely government undertakings, but more often by combinations of public and private funds and of governmental and individual initiative. The relative amount and importance of the contributions from the two sources varied widely; there were even cases in which a single connected line of communication was constructed partly as a public work, where the difficulty was the greatest, partly by mixed enterprise, and perhaps partly by unaided private investment. Government support might take the form of subscriptions to stock, or of loans to the companies or guarantees of their own obligation, or of direct subsidy in cash or kind or credit. Sometimes the governmental claims were given a preferred position as compared to those of private subscribers or lenders; but in other cases the public claim was deliberately subordinated (as in the early days of the Virginia system or on the first transcontinental railroad) in the belief that this would serve more effectively to "elicit individual wealth for public improvement." [38]

By the time the federal government returned to the field of internal improvements in the Era of National Subsidy, one expedient seems to have been ruled out on grounds of accepted political principle. Improvements were not, it appeared, to be constructed by the "direct agency" of the national government. With this exception, the choice of means appears usually to have been made on a quite pragmatic basis. In a period in which public statements on economic policy were so largely cast in terms of individualism and laissez faire, little difficulty was found in justifying government support—and, on occasion, public construction—of the transportation facilities that were so eagerly desired. The overriding problem was to obtain enough capital: from what sources and by what means, were secondary questions. What the United States Supreme Court said of the Pacific Railroad legislation may be taken as applying to the history of internal improvements in general: "The public mind was directed to the end in view, rather than to the particular means of securing it." [39]

It is, however, possible to observe certain general tendencies in the choice of means. As between the different levels of government, there was clearly some disposition to select the one near at hand. The rejection of the federal government as the principal agency of the program is partly to be explained by constitutional scruples and the doctrine of states' rights and perhaps even more, as has been suggested, by the inability of rival local and regional interests to agree on a common national program. Yet the selection of the alternative agencies is also to be understood as reflecting the fact that the individual states and the ambitious commercial cities were themselves objects of political and civil loyalties that were often more intense, and much more easily linked with specific economic interests, than the more general sentiment of national patriotism.

This argument does not wholly apply to the emphasis in the later decades on the role of towns and villages and counties. In most of these cases, reliance on local government aid was not a matter of deliberate preference but rather a final expedient to which the advocates of improvements could turn after state action had been made unconstitutional. Yet the fact that this last resort was available at all, and that it was so extensively used, is itself the reflection of a significant characteristic of nineteenth-century America. In his opinion on a local aid case, an Indiana judge contrasted the United States with other nations and declared that "in this country the political tendencies are centrifugal." [40] That local governments had the authority to take these promotive measures, and that the smaller communities were willing to incur such heavy charges for them, are facts that differentiate the American experience from that of most other countries. What they suggest and illustrate is the importance of the "booster" spirit in American economic development.

A second general tendency appears to have been to give primacy wherever practicable to the private rather than the governmental elements in the combined effort. One sign of this was the preference so generally shown for mixed enterprise, or for governmental aid to private corporations, rather than for public works. State enterprise was the dominant method for several decades of canal building, and there was occasional resort to the expedient of municipal enterprise throughout the nineteenth century and even into the twentieth. In the first flush of New York's success with the Erie Canal, De Witt Clinton often urged its sister states to follow its glorious example in creating their own public works. Yet with this exception, it is hard to find many expressions of preference for direct government action as against mixed enterprise. By far the greater part of public support to railroads was given in the form of subscriptions or other aid to private companies.

In the conduct of mixed enterprise, moreover, there was a general tendency to leave initiative and the responsibility of management mainly to the private elements in the combination, even in cases in which the greater part of the funds came from public sources. Government directors in improvement corporations sometimes performed valuable services in exposing and preventing fraudulent practices. In certain cases the attitude they took was too independent to suit the representatives of private investors. President John W. Garrett of the Baltimore and Ohio, for example, waged a long campaign to rid his company of its majority of state and city directors. But there were a number of cases in which public directors were never appointed, and many others in which governmental representation was rendered largely ineffectual by the lack of interest and ability—and occasionally of courage—shown by the public directors.

The relatively passive role played by government representatives was, however, only partly the result of their own individual shortcomings or of a general laxness in public administration. It was also in part a matter of deliberate legislative policy. Public and individual investors were thought of as sharing the overriding interest in the completion of a proposed improvement, and it was often believed that this common purpose would be better served under private leadership. When Virginia's Board of Public Works expressed its reluctance "to obtrude upon the detailed administration of the various railroad companies," its attitude was entirely consistent with the spirit of the laws under which it was acting. The proposals on which Virginia's system of mixed enterprise was based stated explicitly that the state's control "should extend no farther than to correct or prevent such abuses on the community at large, as might be apprehended from the too eager incentive of gain." The state's participation in stock and voting rights was accordingly limited to two fifths of the whole. Even when the state's subscription was later raised to three fifths, the legislature maintained the philosophy of the original law by holding the voting power of the state's proxy to two fifths instead of assuming majority control.[41] In Pennsylvania, the same policy was embodied in a law providing that the private stockholders should always elect a majority of the board of directors of the Pennsylvania Railroad no matter how large a proportion of its stock might be subscribed by local governments. In a number of Southern and Western states, legislation provided for the distribution of municipal or county stock to individual stockholders, and the practice was sometimes justified on the double ground that it was inexpedient for local governments to attempt to exercise control over the company and that the personal interest of the taxpayers would stimulate them "to a vigilant supervision of the conduct of the work." [42] In the case of the federal land grants, a British investigator reported to the Privy Council that the great merit of the system was that it gave needed assistance without imposing the penalty of interference with management.[43] Many Americans of the time would have accepted his argument, and the lightness of the supervisory hand of government was often regarded as an advantage rather than a disadvantage of the improvements program.

A final illustration of the same tendency is provided by the manner in which public promotion in the various regions came to an end. There were dramatic cases of revulsion against particular programs long before the developmental period was over, but these were often followed by the renewal of aid under different forms and by different agencies. In other areas, the practice of public assistance fell gradually into disuse as demands for such aid grew less frequent and less urgent. Few Americans questioned the principle laid down by the *American Railroad Journal* that public participation should end "so soon as private capital shall be equal with the task of building the roads." [44] In some cases officials of established companies felt that this rule was violated in the extension of aid to later competing lines. But public capital was never at any stage of the movement forced upon an unwilling company or employed on undertakings which private capitalists were eager to exploit without assistance. Government action in this field exhibited no tendency toward indefinite expansion. Instead, it diminished and died out as the railroads reached the "state of progress" in which they needed no further aid and as the "growth of wealth, . . . upheaving of enterprise, and . . . combination of capital . . . made private associations equal to what only a great Commonwealth could have . . . attempted" at an earlier time.[45]

It is, indeed, a striking feature of the American experience that the public and private roles in promotion were almost always thought of as cooperative rather than competitive. Enthusiastic advocates of improvements, like Jared Sparks of the *North American Review*, addressed their appeals "equally to individuals, corporate bodies and state legislatures." [46] Public meetings and conventions organized to build up sentiment for government action also adopted resolutions urging individual subscriptions and donations. Canvassers for the Western Railroad were instructed to call on every householder in Boston, as in a modern drive for for the Red Cross or the Community Fund, and "to urge upon each one, as a matter of duty, as well as of interest, to do their share to advance this great work." "Every man," said an Oregon paper, "can help some. . . . The right of way ought to be cheerfully donated in every case. Crossties can be easily furnished by persons along the line, each furnishing a few, and taking their pay in stock or lands." [47] Even the prohibition of public aid did not always put a stop to collective action on behalf of the railroads. In a number of cases, citizens' committees or boards of trade took over responsibility for raising the sums of money needed to persuade the companies to build through their communities. In 1869, for example, when the decision of an Iowa court prevented Marshalltown from making a public donation to a railroad, the leading citizens closed their offices, put the town band on the streets, and raised the required amount as a voluntary subscription.

Mathew Carey, who supported many improvement companies but declared that he had "never subscribed to a single share with a view to profit," is not to be regarded as the typical stockholder. There were also, as we have seen, cases in which protestations of public purpose were no more than a cover for cynical raids on the public treasury. Yet it would be quite impossible to read the history of American improvements without realizing that they were commonly regarded as a "great cause" appealing to the enthusiasm and "zeal" as well as to the economic interest of subscriber and citizen.[48]

This mingling of public and private action is to be understood in terms of the purpose of the improvements and the way in which they were to be used. They might become, as De Witt Clinton prophesied, monuments of glory to their projectors. They might well serve, as George Washington and many others hoped, to promote the political unity of the country by binding together its various sections. But they could do these things only by opening the way for extended settlement and expanded trade and production and by giving economic opportunities to farmers and planters and speculators, to lumbermen and cattlemen and miners, to merchants, and manufacturers and financiers, for more effective exploitation of the resources of the continent. The opportunities themselves, as everyone understood, were to be seized by private effort. When the Erie Canal was built as a great public undertaking, no one advocated a battery of additional governmental measures to make sure that settlers would move to northern Indiana and Michigan, to instruct them that wheat would be an appropriate crop to raise, to encourage merchants to buy the wheat and millers to turn it into flour, and to induce New England manufacturers to sell their products in the new settlements. Responses such as these were taken for granted and were, in fact, forthcoming in this early case and throughout the history of American improvements. As the movement continued, federal policies of land disposal were changed to promote more rapid settlement and to influence its pattern. Some of the states, as well as the railroads themselves, took steps to encourage and assist immigration, and the federal government and the land-grant colleges began to supply some degree of technical guidance to the farmers. But these and other governmental measures were not intended to substitute centralized planning for individual decisions in the use of the new facilities for transportation. The fundamental assumption remained unchanged. Governmental funds were to be used for undertakings which, because of their size or nature, could not profitably be carried through by unaided private capital, but the opportunities they created were to be exploited by the efforts of individuals and private corporations.

The passage quoted at the beginning of the section suggests that it might appear surprising to find so much government action in a country which could be described at the beginning of the present century as "the place where 'State interference' has played the smallest part, and individual enterprise has been given the largest scope, in industrial affairs." Yet a review of "this remarkable movement toward State enterprise" [49] suggests a different emphasis. If it is recognized that developmental transportation has rarely been supplied in any country by purely private means, the vigorous action of American governments becomes no paradox but merely a reflection of the fact that the United States provided during much of the nineteenth century the world's greatest example of economic development. If attention is directed to the manner in which governments made this contribution, what appears is a set of characteristics long familiar to students of American history. Among them are a decentralized structure of government, a distrust of centralized national plans, and a pragmatic attitude in the choice of means. Students from other countries might find two aspects of the movement extraordinary: One is the degree of authority exercised by local governments and the strength of community or "booster" spirit in the promotion of improvements; the second is the importance of voluntary civic organizations both as pressure groups influencing government decisions and as agencies of direct collective action. Finally, the preference for mixed as against purely public enterprise, the emphasis on private leadership in the joint undertakings, and the reliance on private interests to exploit the opportunities offered by the improvements, show the degree to which this use of the public powers conformed to the traditions, and served the purposes, of individualistic free enterprise.

Under the circumstances of nineteenth-century America, it seems clear that the public promotion of canals and railroads did in fact promote development. With its stimulus the economy grew more rapidly than it would have done if these parts of the nation's overhead capital had been left entirely to private investors to supply. It is no less clear that the effectiveness of the stimulus depended largely on the vigor and promptness with which individuals and business enterprises put these improvements to profitable use.

The full set of conditions which governed the program of internal improvements is not likely to be duplicated in this or any other country. In the United States of the twentieth century, the most conspicuous growth of governmental activity has been in the fields of business regulation, welfare, and national defense. Yet neither government subsidy nor public efforts to promote economic development came to an end with the passing of the frontier and the completion of the railroad network. The St. Lawrence Seaway has recently been completed as an international public work. A number of states have revived the practice, under the old names, of building turnpikes and toll roads. Railroad corporations are again asking and receiving subsidies, basing part of their claim on the federal, state, and local aid that has been given to their rivals in road transport and aviation.[50] The competition of municipalities in the provision of airports recalls their earlier competition in building depots and branch railroads, and state and cities offering inducements to attract manufacturing plants find themselves in much the same bargaining position as those that offered inducements to secure railroad connections.

A book on public promotion in the twentieth century would reverse the proportions of the present one and devote more space to the activities of the federal government than to those of state and local authorities. A number of developmental measures, including the establishment of the Tennessee Valley Authority, owed part of their stimulus to the assumption of federal responsibility for providing employment during the Great Depression. Other significant contributions have been by-products of national defense. It is on this basis that the federal government bore much of the cost of creating the industry of aviation and has so far met all but a small fraction of the cost of developing atomic energy for industrial uses. The new centralization has changed the conditions of public promotion but has by no means eliminated all of its problems. Conflicts of regional interest are still difficult to resolve. Criteria of selection remain hard to establish, though ingenious methods have been devised for comparing estimated costs with "the benefits to whomsoever they may accrue."[51] Public and private funds and efforts are intermingled in a great variety of undertakings—in hydroelectric power, atomic energy, and defense industries, for example, as well as in housing. These modern forms of "partnership" raise problems not unlike those of nineteenth-century mixed enterprise and require no less vigilance and ingenuity to make sure that both their public and private purposes are really served.

The position of the countries today described as "underdeveloped" or "developing" is in some respects more closely analogous to those of the century during which a largely "undeveloped" United States turned itself into a great industrial nation. Most of these countries give a major role in their economic planning to the provision of "public overhead capital," often including roads, railways, waterways, and aviation. If their leaders were to examine the record of the American experience, with the extensive resort to government action that has been described, they would hardly draw the conclusion that they must try to create their own transportation facilities entirely by private means. Most of them, indeed, are likely to place even greater reliance on the agency of the state, partly because "development" has become so overriding a political issue and partly because they will in many cases be unable to count on rapid private responses to economic opportunities and incentives. For this last reason, planners in a number of countries will consider that improvements in transportation are not likely to be effective unless they are accompanied by public planning for complementary developments in agriculture and manufacturing.[52] Few nations are likely to leave as much of the responsibility to local governments as in nineteenth-century America, and no doubt many would prefer the orderliness of a Gallatin Plan to the extraordinary diversity of methods which replaced it. Certainly no nation would wish to follow all the vagaries of the American experience. Yet other governments will encounter some of the same difficulties in evaluating projects, in composing conflicts of local interest, and in devising an adequate machinery of administration. Those who are planning the programs of the developing nations would be fortunate indeed if they could enlist in their effort the civic spirit that led so many American communities to make sacrifices to obtain their canals and railroads. The provision of public overhead capital will not promote development in any country unless the advantages it offers are seized, by private or governmental means, with somewhat the same vigor with which American farmers and merchants and manufacturers made use of the opportunities created by the Erie Canal.

The range of possible expedients is a wide one, as the American experience suggests: differing circumstances require different combinations of private and public effort. Yet for many nations the fundamental problem may be stated in the same terms—to find the ways in which the powers of government can best be used to provide a favorable setting for economic activity and in which the energies of individuals and business enterprises and voluntary organizations can best contribute to the common end of economic development.

NOTES

1. THE SPIRIT OF IMPROVEMENT

1. The quotation at the beginning of the Introduction, from a letter to the Marquis de Chastellux, is one of many expressions of Washington's interest in internal improvements. John Marshall, *The Life of George Washington*, ed. by J. Crissy (Philadelphia: 1850), p. 65.

2. American Philosophical Society, *Minutes*, Vol. I (1769–70).

Annals of Congress, Ninth Congress, Second Session (1807), Vol. II, p. 65.

3. *Report of the President of the North Carolina Railroad to the Governor of the State*, 1859.

4. *A Collection of All Laws and Resolutions of the General Assembly of Virginia Relating to the Board of Public Works*, p. 43–57.

5. Heath, Public Cooperation in Railroad Construction in the Southern United States to 1861 (MS), pp. 59–60.

Lively, "The American System."

6. The significance of the government's scientific and technological contributions to transportation development was pointed out by Forest G. Hill in a Columbia University dissertation in 1950. See Hill, "Government Engineering Aid to Railroads"; also his *Roads, Rails and Waterways*. The importance of military roads as civilian highways to the West is emphasized in Jackson, *Wagon Roads West*.

7. Callender, "The Early Transportation and Banking Enterprises of the States."

8. Adam Smith, *An Inquiry into the Nature and Causes of the Wealth of Nations*, ed. by James E. Thorold Rogers, 2d edition (Oxford: 1880), II, 306, editor's note.

9. "Majority Report of the Committee on the Pacific Railroad," 40th Congress, 3d Session, pp. 10, 19.

10. Thorner, Chap. VII, "Private Enterprise at Public Risk," in *Investment in Empire*.

Jagtiani, *The Role of the State in the Provision of Railways.*

Jenks, *The Migration of British Capital,* pp. 134–52 and 167–70, emphasizes the role of railway contractors and concession-seekers in stimulating government expenditures on railroad building.

11. ARJ IX (1839), 289–93.

12. Cranmer, The New Jersey Canals (MS).

13. Frank W. Fetter, "History of Public Debt in Latin America," *American Economic Review,* XXXVII (1947), 147–48.

14. Memorial of the State of Alabama, February 4, 1872.

15. *Annals of Congress,* 31st Congress, 1st Session, p. 851.

16. ARJ XXXIII (1860), 100; see also XXV (1852), 26; XXVIII (1855), 281, 353, 514, 609–10, 673.

17. Kirkland, *Men, Cities and Transportation,* I, 111–12, 161–63.

18. The London *Times,* March 30, 1874, quoted in Oscar O. Winther, "Promoting the American West in England, 1865–1890," *Journal of Economic History,* XVI (1956), 513.

19. Brayer, *William Blackmore,* II, 2, 58.

20. Hungerford, *The Story of the Baltimore and Ohio Railroad,* I, 55.

21. Sanderlin, *The Great National Project,* p. 57.

22. Act of March 3, 1873.

23. In addition to many cases of state and local exemptions, the federal government sometimes remitted duties on imported iron. By an act of March 27, 1787, the state of South Carolina offered credits against duties on the importation of slaves as encouragement to the improvement of the Catawba and Wateree rivers.

2. AN ERA OF NATIONAL PROJECTS

1. "Report on Roads and Canals," presented to the Senate by the Secretary of the Treasury, April 4, 1808, *American State Papers,* Miscellaneous, I, 724–921. The Report proper is on pp. 724–41; the rest is Appendix. Later citations will be to Gallatin's Report.

2. *Annals,* 4th Congress, 2d Session, pp. 2052–65; 2d Congress, 1st Session, pp. 229–41. On two occasions the usually impersonal recorder of the *Annals* made uncomplimentary references to the quality of the debates on post roads: 4th Congress, 2d Session, p. 2059; 9th Congress, 2d Session, pp. 623–24.

3. *Annals,* 5th Congress, 1st Session, pp. 621–22.

4. Instances are given in Marcuse, Local Public Finance in Colonial Connecticut (MS).

5. *Annals*, 31st Congress, 1st Session, p. 851.

6. The Society for Promoting the Improvement of Roads and Inland Navigation, quoted in Gallatin's Report, p. 860.

7. Davis, "Toll-Bridge and Turnpike Companies" in his *Essays in the Earlier History of American Corporations*, Vol. II, Chap. IV. Durrenberger, *Turnpikes*, esp. p. 5 and p. 98, note 6.

Gallatin's Report, pp. 866–910.

8. Gallatin's Report, esp. pp. 860, 811, 730, 727, 802.

9. *Ibid.*, pp. 765–88, 863–65, 736, 823–26.

MacGill, *History of Transportation in the United States before 1860*, p. 213.

10. Gallatin's Report, p. 730.

11. *Ibid.*, pp. 865, 791, 818, 819, 766, 829. On the management of the Western Inland Lock Navigation Company, see Nathan Miller, "Private Enterprise in Inland Navigation," pp. 398–413.

12. Gallatin's Report, pp. 757–58, 878, 827.

13. *Annals*, 4th Congress, 1st Session, pp. 65–66, 76, 1342; 4th Congress, 2d Session, Appendix, pp. 2904–5.

Clement L. Martzolff, "Zane's Trace," *Ohio Archaeological and Historical Society Publications*, XIII (1904), 287–331.

Jordan, *The National Road*, Chap. III, "Ebenezer Hews a Trace."

14. *The Writings of Albert Gallatin*, ed. by Henry Adams (Philadelphia: 1879), pp. 76–79. Cited in Jordan, *The National Road*, pp. 71–72. The date is February 13, 1802.

15. *American State Papers*, Miscellaneous, I, 432–34. Quoted in T. B. Searight, *The Old Pike*, (Uniontown, Pa.: 1894), pp. 21–24, 25–27.

Annals, 9th Congress, 1st Session, pp. 22–25, 835–40.

Jordan, *The National Road*, p. 74.

16. *American State Papers*, Miscellaneous, I, 474–77.

Searight, *The Old Pike*, pp. 28–35.

17. *American State Papers*, Miscellaneous I, 714–15, 914. Searight, *The Old Pike*, pp. 37–42. For Jefferson's attitude, see Jordan, *The National Road*, pp. 79–80; and Joseph Hobson Harrison, The Internal Improvement Issue in the Politics of the Union, 1783–1825 (MS), pp. 248–49.

18. *American State Papers*, Miscellaneous, I, pp. 419–52, 453–56, 479.

19. *Ibid.*, pp. 454–55.

20. Everett Somerville Brown (ed.), *William Plumer's Memorandum*

of Proceedings in the United States Senate, 1803–1807, (New York: 1923), pp. 628–29. See also John Quincy Adams, *Memoirs,* I, 460–61 and *Annals,* 9th Congress, 2d Session, pp. 33–35, 56–60, 80–87. The text of the speech has not been preserved.

21. *Annals,* 9th Congress, 2d Session, pp. 77–78. Letter of February 2, 1837, quoted in Brooks Adams's introduction to Henry Adams, *The Degradation of the Democratic Dogma,* pp. 24–25.

22. For five rollcalls on which Adams and Worthington were opposed on internal improvement issues, see *Annals,* 8th Congress, 1st Session, pp. 273, 281; 2d Session, p. 37; 9th Congress, 2d Session, pp. 95–96.

"Had Adams died within the next decade, he would probably have been remembered by his colleagues . . . as the dogged opponent of internal improvements—rather than as their foremost friend." Harrison, Internal Improvement Issue (MS), p. 199; see also pp. 196–99 and p. 181. But in 1823 Adams was toasted as "the first to elicit in the Senate of the United States the luminous report of Albert Gallatin upon internal improvements." *Ibid.,* p. 550.

23. *Annals,* 9th Congress, 2d Session, pp. 88–90, 95–97.

24. In this, Gallatin exhibits "that zest which characterized all the Jeffersonian circle—the delight that, physically as in other ways, their America was a new world. The map, for him, if not so malleable as sculptor's clay—for it already possessed a structure that must be respected—was nevertheless a thing to manipulate. The natural lines of communication must be developed, reinforced, articulated with a delicacy and precision worthy of the son of a people who made watches." Harrison, Internal Improvement Issue (MS), p. 223.

25. *Annals,* 11th Congress, 1st Session, pp. 522–25, 613, 1385–1401, 1443.

26. Both bills: March 29, 1810, January 12, 1811. Chesapeake and Delaware alone: December 20, 1813.

27. *Report of the Commissioners* (Albany: 1812), pp. 32–38.

28. *Annals,* 12th Congress, 1st Session, pp. 1078–80. Gallatin's famous biographer describes him as throwing behind him "all his old hopes and ambitions, all schemes for discharging debt and creating canals, roads, and universities" and training "his energies to the single point of defending the Treasury and resisting follies." Henry Adams, *Life of Albert Gallatin,* p. 399. The same passage says that he regarded the "habit of borrowing money with horror" and as a last resort "to be reserved for war." In his committee testimony, however, Gallatin ex-

pressed approval of the proposal to borrow for internal improvements on the security of the public lands.

29. *Annals*, 14th Congress, 2d Session, pp. 851–54.

30. *Ibid.*, pp. 296, 851–70, 874–914, 916–23, 166–80, 185–88, 190–91, 1051–52.

31. This was true of the amendments proposed by Senator James Barbour of Virginia in 1817, by Senator Van Buren of New York in 1824, and by Congressman Bailey of Massachusetts in 1825. *Annals*, 15th Congress, 1st Session, pp. 21–24; 18th Congress, 1st Session, pp. 134–36; *Register of Debates in Congress*, 19th Congress, 1st Session. On the other hand, Senator Samuel Smith of Maryland proposed in 1823 an amendment to empower Congress "to adopt and execute a system of internal improvements, confined to great national purposes." *Annals*, 17th Congress, 2d Session, pp. 200, 227, 290.

Though Jefferson, Madison, and Monroe all urged the adoption of amendments to clarify the government's powers, these proposals were opposed not only by anti-improvement men but also by pro-improvement men who were convinced that the government had ample power to act under the constitution as it was.

32. Sadove's valuable study, Transport Improvement and the Appalachian Barrier (MS), concludes that the Cumberland Road and its rival the Pittsburgh Pike had an important economic effect in maintaining the east-west route for the supply of manufactured goods to the Middle West against the competition of steamboats on the Mississippi but that they did not displace the river as the main outlet for the heavy products of the West. See especially pp. 309, 420. The Road thus met a part but not the whole of the test which Gallatin had laid down for it "as a national object." *American State Papers*, Miscellaneous, I, 718.

33. Jordan, *The National Road*, p. 159.

34. "Report on Roads and Canals," in Calhoun, *Works*, V, 40–54; *Annals*, 17th Congress, 1st Session, pp. 606–10; 18th Congress, 1st Session, pp. 1013–42 and Appendix, p. 3217.

35. James D. Richardson, *A Compilation of the Messages and Papers of the Presidents* (Washington: 1909), II, 305–16. A recent historian notes that these passages went further than Henry Clay thought wise and says that they represented "an extraordinary, perhaps a unique, example of the art of suicide by manifesto." George Dangerfield, *The Era of Good Feelings* (New York: Harcourt, Brace, 1952), p. 148. But the course of legislation showed that there was substantial political backing for particular schemes of internal improvement if not for

Adams's broader philosophy. A serious attempt to support a great system of improvements from revenue from land sales would have confronted the mounting pressure for disposing of the land to settlers on progressively easier terms.

36. *Register of Debates in Congress,* 19th Congress, 1st Session, pp. 2551–52.

37. "Nature has already made this connexion nearly complete. . . . A great circumnavigation might be formed . . . connecting in its course, by navigable waters, the whole of the western and Atlantic countries." Congressman Paul B. Porter, *Annals,* 11th Congress, 1st Session, pp. 1390–92.

38. For the Ohio road, a quantity of land equal to half of a strip two sections wide on the western side of the road; for each of the canals, a quantity of land equal to half of a strip five sections wide on each side of the canal. The allotment to the Indiana road consisted of land just obtained by an Indian treaty.

39. *Annals,* 14th Congress, 1st Session, p. 110.

40. The Maryland subscription was $500,000 and those of the three District of Columbia cities totaled $1,500,000. The latter could, of course, act only with congressional authorization. It has therefore been argued that the "financial help . . . received from the Federal Government must . . . be reckoned not at one million dollars but at two and a half million." Ward, *The Early Development of the Chesapeake and Ohio Canal Project,* pp. 85–87.

41. John Quincy Adams, *Memoirs,* VIII, 48–50.

42. Richardson, *Messages and Papers of the Presidents,* II, pp. 483–94.

43. The administration did not send a representative to vote its stock at the 1829 meeting but it did participate in the 1832 meeting in the hope of removing a political opponent from the presidency of the company. In 1830, the House Committee on Internal Improvements recommended against further aid until the relative merits of railroads and canals had been determined. Sanderlin, *The Great National Project,* p. 101.

44. "Farewell Address of Andrew Jackson to the People of the United States," Washington, 1837, quoted in Joseph L. Blau (ed.), *Social Theories of Jacksonian Democracy* (New York: Hafner, 1947), p. 305.

45. Searight, *The Old Pike,* Chap. XIV, tabulates the appropriations. The last substantial amount was voted in1838.

46. By counting all these items, one author has reached the conclusion that expenditures for internal improvements increased from an annual average of $702,000 under John Quincy Adams to $1,323,000 under Jackson. Van Deusen, *Economic Bases of Disunion in South Carolina,* p. 128. See also Taylor, *The Transportation Revolution, 1815–1860,* pp. 20–21. With respect to rivers and harbors improvements, Presidents Van Buren, Tyler, Polk, Pierce, and Buchanan all followed a stricter interpretation of the Constitution than Jackson. Hill, *Roads, Rails and Waterways,* Chaps. VI and VII, esp. p. 196.

47. "Until 1840 or 1850 [West Point] was practically the only school in America offering systematic instruction in any form of engineering." The great contribution of its graduates to the early railroad industry was made partly in the service of the Army Engineers, partly on furloughs, and partly after resignation from the Corps. Hill, *Roads, Rails and Waterways,* Chaps. IV and V, esp. p. 143.

48. Congressman George Holcombe of New Jersey. *Annals,* 18th Congress, 1st Session, pp. 1013–21, quoted in Harrison, Internal Improvement Issue (MS), p. 560.

49. Senator Nathaniel Macon of North Carolina. *Annals,* 15th Congress, 1st Session, p. 179.

50. For condemnations of "pernicious copartnerships," see *Register of Debates,* 18th Congress, 2d Session, pp. 301, 676, 1613–16; 19th Congress, 2d Session, p. 1387; 20th Congress, 1st Session, p. 2616. For an almost unique instance of an argument that the rights of existing corporations would be impaired by federal competition see *Annals,* 15th Congress, 1st Session, p. 1358.

51. John Randolph of Virginia and Thomas W. Cobb of Georgia. *Annals,* 15th Congress, 1st Session, p. 179; 2d Session, pp. 657–58.

52. Richardson, *Messages and Papers of the Presidents,* II, 516.

53. *North American Review,* XII (January, 1821), 17–20; XIII (July, 1821), 144. The first article is attributed to Jared Sparks, the second to John L. Sullivan.

54. For example, Austin of Virginia and Robertson of Louisiana. *Annals,* 14th Congress, 2d Session, p. 866; 15th Congress, 1st Session, p. 1210. The Erie-Cumberland comparison was made by Congressman T. R. Mitchell of South Carolina on February 16, 1829. Jordan, *The National Road,* pp. 164–65.

55. Thomas Bayly, cited in Harrison, Internal Improvement Issue (MS), p. 517. Robert Wright. *Annals,* 14th Congress, 2d Session, p. 921.

56. *Annals,* 11th Congress, 2d Session, p. 1400. See also speeches

of Henry Clay and James Breckinridge of Kentucky and Andrew Stewart of Pennsylvania. *Ibid.,* 15th Congress, 1st Session, p. 1177; 17th Congress, 1st Session, pp. 1478, 1485. But Western opinion, as Jackson's attitude demonstrates, was not unanimous in favor of federal action.

57. Harrison, Internal Improvement Issue (MS), p. 479, refers to "New York's discontent—in which the Virginia Dynasty's refusal to aid the Erie Canal was a prime ingredient." It might be of interest to speculate as to what the fate of national internal improvements might have been if Clinton and Morris had managed to secure sufficient votes for their omnibus measure or if the Bonus Bill, passed largely by New York and Pennsylvania votes, had not been vetoed in the year in which New York State decided to build the Erie Canal with its own resources.

58. Quoted by Brooks Adams in his introduction to Henry Adams, *Degradation of the Democratic Dogma,* pp. 24–25.

59. Alexander Hamilton, "Report on Manufactures," in his *Works,* ed. by Henry Cabot Lodge (2d ed., New York, Putnam, 1903), IV, 161.

60. The argument of this and the preceding eight paragraphs was presented at somewhat greater length in Goodrich, "National Planning of Internal Improvements," pp. 36–43. See also Taylor, *The Transportation Revolution, 1815–1860,* pp. 20–21, which reaches the conclusion that: "The real obstacle which defeated a national system of internal improvements is to be found in the bitter state and sectional jealousies which were wracking the new nation." Harrison, Internal Improvements Issue (MS), gives a fresh, detailed and sympathetic discussion of the constitutional issues.

61. Brooks Adams, in his introduction to Henry Adams, *Degradation of the Democratic Dogma,* pp. 24–25.

3. THE APPALACHIAN COMPETITION

1. *A Collection of All Laws and Resolutions of the General Assembly of Virginia Relating to the Board of Public Works,* p. 21.

2. Whitford, *History of the Canal System of the State of New York,* I, 59, 61. The argument for the direct route was stated in a resolution adopted in 1808 on the motion of Joshua Forman: "And whereas the state of New York, holding the first commercial rank in the United States, possesses within herself the best route of communication between the Atlantic and western waters, by means of a canal between the tide waters of the Hudson river and Lake Erie, thro' which the wealth and trade of that large portion of the union, bordering on the upper

lakes, would forever flow to our great commercial emporium." *Assembly Journal,* 1808, p. 58.

3. General Tallmadge in the Assembly of 1824. Whitford, *History of the Canal System,* p. 121.

4. The selection of revenues was made in part at least on a theory of local benefit, the salt tax being thought of as a contribution of the West and the auction duties of New York City. But the legislators rejected a proposed property tax on cities, villages, towns, and counties immediately benefiting from the canal; and a provision authorizing a special tax on land within twenty-five miles of the canal was never utilized. Grants and donations could also be received.

5. Whitford, *History of the Canal System,* I, 125.

6. Miller, Enterprise of a Free People (MS). Miller points out that the New York Bank of Savings purchased and held as an investment a considerable amount of the earlier issues.

7. Segal, Canal Cycles (MS), pp. 16–17 and Table 3, p. 15. The cumulated surplus revenue was $8,228,439, but "cumulated subsidies" accounted for 51.4 percent of this, or $4,232,439. These included $260,-413 paid out of the general fund for operating deficits on the lateral canals and $305,938 from sales of canal land in addition to the proceeds of taxation.

8. Note esp. Whitford, *History of the Canal System,* Vol. I, Chap. XXV, "The Influence of the Erie Canal, "and Sadove, "Isolating the Economic Effects of the Erie Canal," in Transport Improvement and the Applachian Barrier (MS), pp. 162ff. "The Erie Canal," says Sadove (p. 180), "provided an export route for the products of the northern division, whereas the southern products still went through New Orleans." The dividing line crossed Ohio, Indiana, and Illinois approximately along the route of the National Road.

9. A memorial from a meeting held at Utica in 1835. Whitford, *History of the Canal System,* I, 143–44.

10. "The Empire State does not as yet condescend to enter into competition with her subjects." ARJ IX (1839) 289–93.

After 1848, freight could be carried in any season. In 1851, the railroad tolls were abolished. Segal, Canal Cycles (MS), n. 181, pp. 91–92.

11. Pierce, *The Railroads of New York,* esp. Table 2, p. 193, "State Aid to Railroads." The total of $2,095,512 for the ten railroads includes $638,301 which the state paid as interest on the securities of three defaulting companies. The repayment on the six lines amounted to $770,000.

Reubens, Government Financing of Private Enterprise, now in

progress, will consider these as well as other cases in which New York State financed or invested in the operations of private corporations or individuals.

12. Quoted in Lincoln, *The Constitutional History of New York,* II, 215. The year was 1846.

13. "During the pressure we have experienced, the industry of the citizens has been stimulated, and the wages of labor, the price of the products of the earth, and the value of property have been sustained in the prosecution of this system." Whitford, *History of the Canal System,* I, 160–61, quoting the Governor's message to the 1840 legislature. See also Segal, Canal Cycles (MS), pp. 49–51.

14. Pierce, *The Railroads of New York,* Table I, pp. 178–92, "Municipal Aid to Railroads"; Chap. II, "A Municipally Owned Railroad"; and pp. 116–18. There were two cases of local aid, in small amounts, in 1837. Albany's aid to the Mohawk and Hudson consisted of a loan of $125,000, which was repaid; an endorsement of $100,000 of the company's bonds, which the city was never called upon to honor; and a donation of $25,000. Pierce describes the Troy Union episode (pp. 117–18) as "an instance of community intimidation by a powerful corporation."

15. Segal, Canal Cycles (MS), Table 1, p. 11; Appendix Table 1, p. 117, gives construction costs, 1834–61 as $47,722,000, excluding the overhead and administrative costs of the Canal Commissioners. Whitford gives gross construction costs, 1817–33, as $9,425,000. If, as Segal suggests, $400,000 is subtracted from the Whitford figure to allow for administrative costs, the total net figure for construction becomes $56,747,000. If, on the other hand, the administrative costs were in both cases attributed to construction, the total would be about $60,500,000.

The approximate net income, after interest, for the period 1817 through September 30, 1861, is derived as follows:

Income		
Tolls	$76,300,000	
Other	5,100,000	
Total		$81,400,000
Expenditures		
Interest	30,400,000	
Other	23,600,000	
Total		54,000,000
Net Income		$27,400,000

The basic data are from State of New York, "Annual Financial Report of the Auditor of the Canal Department," *Assembly Document No. 4,* 1863, Statement #81, opp. p. 132. Subsidies from and payments to the General Fund of the state are omitted. The reported totals of expenditures are corrected by (1) adding $3.8 million for the expenses of the Canal Commissioners and (2) subtracting $2.1 for construction paid for out of current account.

Railroad tolls contributed $662,000 to the receipts. A part of the item of interest earned, included in other income, should strictly, as Segal points out, be attributed to the taxes collected for the Canal Fund.

16. From the city of Albany, for the Mohawk and Hudson extending from Albany to Schenectady. The Tonawanda received state aid but applied it to a section that did not become part of the main line. A. F. Harlow, *The Road of the Century; the Story of the New York Central* (New York, Creative Age Press, 1947), p. 57. Pierce, *The Railroads of New York,* gives a number of cases of public aid to roads that became parts of the New York Central system.

17. A contemporary English observer, whose chief interest was in the soundness of the security behind the loans in which Europeans had invested, noted in 1839 the great advantage of New York's policy of providing supplementary tax revenues as part of the basis of the canal fund. "It is by this judicious policy alone that the Erie canal is at this moment the unburdened source of a large revenue to the state, enabling the legislature, without embarrassment, to act in the true spirit of its object, by reducing the tolls when the interests of trade require it, unfettered, by any considerations of a mere fiscal nature. Without the adoption of such a system, a state will not only fail to derive this benefit, but must fail likewise to inspire capitalists with that full conviction of their security which will induce them to come forward on liberal terms to its assistance." Alexander Trotter, *Observations on the Financial Position and Credit of Such of the States of the North American Union as Have Contracted Public Debts,* pp. 83–84.

18. The Chenango Canal was authorized by the legislature in 1833 after the Canal Commissioners had reported in 1829 a negative answer to the question whether the tolls for the first ten years, including the contribution of its traffic to the tolls on the Erie, would equal maintenance and interest. This was one of several occasions on which advocates of particular canals countered the Commissioner's estimates with much more optimistic figures of their own. Whitford, *History of the Canal System,* pp. 676–78 (Chenango); 528–29 (Black River); 614–

15 (Chemung). But Segal gives a number of instances of characteristic *under*-estimates of construction costs on the part of the Commissioners.

19. A partial exception was an Assembly resolution of 1835 asking the Canal Commissioners to prepare a report on the comparative advantages of canals and railroads. Their report, prepared by three engineers, came to the conclusion that canals had in general greater advantages. *Assembly Document No. 296,* 1835. Whitford, *History of the Canal System,* I, 142–45.

Proposals for the establishment of a Board of Internal Improvements, as an agency for coordination, were made and defeated in 1822, 1825 and 1839. Lincoln, *The Constitutional History of New York,* II, 78. Reubens, Government Financing of Private Enterprise (MS), emphasizes both lack of coordination and conflict between the Erie "Canal Interest" and the claims of other areas.

20. Whitford, *History of the Canal System,* I, 129–30.

Jenks, *The Migration of British Capital,* p. 74, notes both this "revelation" and "the marvel that Erie paid its way."

21. Gallatin's Report, Appendix, pp. 834–43.

22. Sadove, Transport Improvement and the Appalachian Barrier (MS), considers it probable that the two roads carried at least 30 to 40,000 tons a year during the eighteen-twenties, as against perhaps 100,000 tons shipped downriver to New Orleans. Since the road traffic consisted mainly of westbound manufactured goods, Sadove suggests that it may have equaled the river traffic in value.

23. Bishop, "The State Works of Pennsylvania," Chap. I. Livingood, *The Philadelphia-Baltimore Trade Rivalry,* pp. 90–91. Hartz, *Economic Policy and Democratic Thought,* pp. 83–85. The state had also guaranteed interest on certain subscriptions to the Union Canal Company.

24. Bishop, "The State Works of Pennsylvania," Chap. II.

25. Rubin, Imitation by Canal or Innovation by Railroad (MS), which contains a detailed account of the Pennsylvania decision. Mathew Carey was the most active leader in the Society, and the architect William Strickland made its investigations in Europe. As Rubin points out, the Society carried its partisanship to the point of suppressing passages in the Strickland report that it considered unduly favorable to railroads. See also Carey, *Brief View of the System of Improvements* and Strickland, *Report on Canals, Railways and Other Subjects.*

26. Bishop, "The State Works of Pennsylvania," pp. 266–77, reproduces the text of the Harrisburg resolutions and the acts of 1826 and

1827. Page 202 quotes "a freeholder of Franklin county" as writing in 1829 that "the Commonwealth, in place of one canal, was, by the log-rolling, rolled into three or more." See also Hartz, *Economic Policy and Democratic Thought*, pp. 131–42.

27. Segal, Canal Cycles (MS), pp. 136–66. Bishop, "State Works of Pennsylvania," Chap. III.

28. Revenue figures are from Bishop, "The State Works of Pennsylvania," Appendix VI, pp. 277–86. Segal, Canal Cycles (MS), p. 215, presents a table of "Net Operating Revenues as Percentages of Original Costs" for the public works as a whole, for the Main Line, for the Philadelphia and Columbia R.R. and for the Delaware Canal. Since official figures of tonnage were not kept on the Pennsylvania system, the comparison relies on the estimates of Michel Chevalier, *Histoire et description des voies des communications aux Etats-Unis* (Paris, 1840), cited in Segal, Canal Cycles (MS), p. 148*n*.

29. Walter Buckingham Smith, *Economic Aspects of the Second Bank of the United States* (Cambridge: Harvard University Press, 1953), Chap. II, esp. p. 179. The same author notes (p. 255) that: "Over $6,000,000 . . . of Pennsylvania issues acquired in accordance with the provisions of the 1836 charter . . . were ultimately sold at an average loss of 47 percent."

30. Segal, Canal Cycles (MS), pp. 174–84. Bishop, "The State Works of Pennsylvania," pp. 222–23.

31. "I never meet a Pennsylvanian at a London dinner without feeling a disposition to seize and divide him; to allot his beaver to one sufferer and his coat to another." Letter of the Rev. Sydney Smith to the London *Times*, November 4, 1843, quoted in McGrane, *Foreign Bondholders and American State Debts*, pp. 59–60. More than half of the state's securities were held in England.

32. Hartz, *Economic Policy and Democratic Thought*, pp. 144–65. Pages 161–80 deal with "The Struggle over Liquidation" and "The End of the Public Works." The popular vote on the 1844 referendum is analyzed on p. 163. Both sales measures were the results of long legislative struggles, with a number of close votes in each house. See particularly the protest of thirteen senators against the sale of the Main Line, Senate *Journal* (1857), pp. 1085–87, and the minority and majority reports of the Committee on Ways and Means on the sale to the Sunbury and Erie, House *Journal* (1858), pp. 357–60, 406–12. The last of these documents makes it clear that one purpose of the sale to the Sunbury and Erie was to facilitate completion of the railroad.

33. Bishop, "The State Works of Pennsylvania," p. 229, and Appendix VII, pp. 287–88.

34. *Harrisburg Weekly Telegraph,* August 1, 1857, quoted in Hartz, *Economic Policy and Democratic Thought,* p. 171. On p. 121, Hartz describes the Pennsylvania Railroad as being, though mixed in ownership, "an outstanding example of the new strength of the transportation corporation."

35. "Government bodies" had paid in $6,750,000 of a total of $12,-355,000. Burgess and Kennedy, *Centennial History of the Pennsylvania Railroad Company,* p. 71. In addition, much of the investment by public authorities in other cases was in roads which were later absorbed by the Pennsylvania.

36. Bishop, "The State Works of Pennsylvania," p. 221. Hartz, *Economic Policy and Democratic Thought,* pp. 83–88. Hartz cites six laws authorizing county subscriptions to turnpike companies, between 1829 and 1837, and notes that subscriptions were sometimes made without such authorization. In 1834 and 1835, the state guaranteed interest on $300,000 for the Danville and Pottsville Railroad and on $350,000 for two navigation companies. Pennsylvania, *Auditor General's Report* (1842), pp. 69–70.

37. Calculated from "Report of Commissioners for the State Stocks," House of Representatives *Journal* (1844), Vol. II, Doc. 7. Bank stock of over $2,000,000, which unlike the improvement stock had yielded substantial dividends, was sold at the same time at a little less than fifty cents to the dollar.

38. Figures of local government investment supplied by Segal. The court case was *Sharpless et al* v. *Mayor &c of Philadelphia,* 21 Pa. St. Rep. 147 (1853). For analysis of the issues see Hartz. *Economic Policy and Democratic Thought,* pp. 86–89, 113–22. The Philadelphia subscription to the Hemphill Railroad was authorized only after the Councils had ordered publication in the newspapers of an answer prepared by the president of the railroad to a series of questions propounded by the Finance Committee of the Select Council. *Journal of the Select Council,* 1852–53, Appendix 36, pp. 144–55. See also Burgess and Kennedy, *Centennial History of the Pennsylvania Railroad Company,* pp. 146–56.

39. Poor, *Manual of the Railroads of the United States for 1868–69,* pp. 20–21. This gives the figure for 1845 as 798 and for 1858 as 2,340. Later references to this work will be to Poor's *Manual.*

40. Hartz, *Economic Policy and Democratic Thought,* Part II, "The

State as Promoter and Entrepreneur." Figures by counties of the popular vote on the anti-investment amendments are given on p. 125. See also Goodrich, "Revulsion against Internal Improvements."

41. Stocks held in 1860: bridge, $9,000; turnpike, $1,230,367; canal and navigation, $325,532; railroad, $181,647; total, $1,746,546. Dividends received: $306. Hartz, *Economic Policy and Democratic Thoughts*, pp. 85–86, 89, 92.

42. Bishop, "State Works of Pennsylvania," Chap. V. Hartz, *Economic Policy and Democratic Thought*, pp. 148–60.

43. Rubin, Imitation by Canal or Innovation by Railroad (MS), p. 6. Table I, p. 29, contains the following comparison between the Erie Canal as constructed and the Pennsylvania Main Line as estimated on the basis of a canal covering the full distance:

	Length (in miles)	*Total Rise and Fall (in feet)*	*Locks*
Erie Canal	363	675–689	83–84
Pennsylvania Main Line	395	3,336	417

As actually constructed, with two sections provided by railroads, the Main Line had 174 locks. Mileages are from Albany to Buffalo and from Philadelphia to Pittsburgh respectively. The distance from Philadelphia to the Ohio River was therefore considerably less than from New York City to Lake Erie.

44. J. R. Tyson, *Letters on the Resources and Commerce of Philadelphia* (Philadelphia: 1852), quoted in Bishop, "State Works of Pennsylvania," pp. 246–47. But note the ingenious attempt to overcome the difficulties of transshipment by demountable "section boats," anticipating modern "piggy-back" devices. House of Representatives *Journal* (1844), Vol. II, Doc. 3.

45. Pennsylvania *Laws*, 1825–26, pp. 344; 1848, p. 274.

46. *Ibid.*, 1853, pp. 218, 609; 1854, pp. 504–05. For lack of representation, see Hartz, *Economic Policy and Democratic Thought*, pp. 96–106. "On the local level," he says (p. 96), "the control factor was taken with varying degrees of seriousness." These variations have not been fully explored.

47. Philadelphia, *Journal of the Select Council*, 1852–53, pp. 109, 121–22. Sunbury and Erie Railroad, *Report of the President and Managers*, February 13, 1854. Pennsylvania Railroad Company, *Annual Report*, No. 6, February 7, 1855, "Minutes of Annual Meeting." Pennsylvania House of Representatives, *Journal* (1853), pp. 490–91, "Petition from Commissioners of Allegheny County." In 1852, a similar sub-

scription to the stock of the Ohio and Pennsylvania Railroad was deferred for several months, after a "hostile vote of the city corporation," until a conference could be arranged between a committee of stockholders and the city councils. Pennsylvania Railroad Company, *Annual Report,* No. 5, February 2, 1852, "Minutes of Annual Meeting," p. 9.

The Allegheny County Cimmissioner who made the protest against the Marietta and Cincinnati subscription was himself also the president of another western connection, the Ohio and Pennsylvania. In the event, the M. & C. subscription failed in its purpose, while the O. & P. became part of the Pennsylvania's line to Chicago.

48. An amendment to make this representation perpetual, irrespective of continued stock ownership by the local authorities, was defeated in the Senate by a vote of 13 to 14. Senate *Journal* (1857), pp. 1008–9, 1085–87.

49. This is suggested by Harvey H. Segal in an unpublished MS, Maryland: A Mixed Enterprise System, of which I have made extensive use.

50. This is reflected in the title of the principal work on the canal: Sanderlin, *The Great National Project.* See also Ward, *The Early Development of the Chesapeake and Ohio Canal Project,* and the earlier works cited by Sanderlin and Ward. For the work of the Corps of Engineers, see Hill, *Roads, Rails and Waterways,* Chap. II.

51. Ward, *The Early Development of the Chesapeake and Ohio Canal Project,* p. 8.

52. Sanderlin, *The Great National Project,* Chap. V. On p. 100, Sanderlin cities a committee report to the Maryland Senate which describes the railroad but not the canal as "decidedly and unqualifiedly a Maryland work."

53. *Ibid.,* p. 123.

54. Rubin, Imitation by Canal or Innovation by Railroad (MS), emphasizes the boldness, or even recklessness, of the decision as related to the knowledge available at the time.

55. Chandler, *Henry Varnum Poor,* p. 38. See also Reisenstein, "The Economic History of the Baltimore and Ohio Railroad, 1827–1853," and Hungerford, *The Story of the Baltimore and Ohio Railroad, 1827–1927.* The experiment with sail power is described in Hungerford, I, 80–81.

56. Goodrich and Segal, "Baltimore's Aid to Railroads," pp. 8–9. The use of the railroad notes is fully analyzed in the Segal manuscript, which gives their total as $1,449,051 as against the Chesapeake and Ohio Company's $1,432,510. These two sets of notes provided the chief

circulating medium of the area during the period in which the Maryland banks suspended specie payment.

57. *Debates and Proceedings of the Maryland Reform Convention to Revise the State Constitution,* esp. I, 132.

58. Sanderlin, *The Great National Project,* Chaps. VIII and IX and Appendix Tables IV and V. The act of 1845 contained the condition that a guarantee of tonnage be obtained from the coal operators who were expected to provide the major source of traffic. Alexandria did not join the other cities in a direct guarantee but instead agreed to indemnify any of its citizens who guaranteed the bonds. *Ibid.,* p. 154.

59. Hidy, *The House of Baring in American Trade and Finance,* pp. 411, 592.

60. Aid realized from Baltimore's subscription of $3,500,000 was $3,491,756 and from Maryland's subscription of $4,000,000 was $3,-353,000. If Wheeling's $500,000, which was guaranteed by the state of Virginia, is taken at face value, the total is $7,344,756 as against an estimated construction cost of $7,623,605 to Cumberland and of about $6,600,000 from Cumberland making a total of $14,223,805. Segal, Maryland (MS).

61. The Northwestern bonds sold at a premium. The B. & O. realized $4,326,000 on the $5,000,000 loan, of which $500,000 was retained for a sinking fund; and the Pittsburgh and Connellsville, $901,355 on the $1,000,000 loan. The 1859 appropriation for the Northwestern was $282,500, but the amount was later reduced to $135,384.

62. Hanna, "A Financial History of Maryland," pp. 76–78.

63. Livingood, *The Philadelphia-Baltimore Trade Rivalry,* p. 124, note 21. Chap. VI is entitled, "Baltimore Reaches Northward."

64. Hanna, "A Financial History of Maryland," pp. 80–81.

65. *Niles' Weekly Register,* L, 65–67. The House Committee on Internal Improvements, which was opposed to the measure, protested that "a work unimportant in itself cannot acquire . . . any of the importance of other works with which it may be associated in a bill or report." "A Citizen of Maryland," *A Short History of the Debt of Maryland* (Baltimore: 1845), p. 37. Both citations are from Segal, Maryland (MS).

66. Livingood, *The Philadelphia-Baltimore Trade Rivalry,* p. 118. The year is 1827.

67. Goodrich and Segal, "Baltimore's Aid to Railroads." The quotation is from p. 17. Receipts and expenditures are given in Table II, pp. 12–13. When the Council opposed the management's proposal to issue an extra dividend in convertible stock in 1856, three of the city's

eight directors and the state representatives voted against the Council's policy. *Ibid.*, pp. 28–29 and note 61.

68. In a letter to Governor Benjamin Harrison, October 10, 1784. *The Writings of George Washington,* ed. by George Washington Ford (New York: Putnam, 1891), X, 404–14, cited in Dunaway, *History of the James River and Kanawha Company,* p. 13. This book is the principal authority on the James River project.

69. Presented December 26, 1812. Reprinted in *A Collection of All Laws and Resolutions of the General Assembly of Virginia Relating to the Board of Public Works.* The quotations are from pp. 37, 33, 38–39.

70. *The Writings of Thomas Jefferson,* ed. by Paul Ford (New York: Putnam, 1895), III, 475.

71. *A Collection of All Laws and Resolutions of the General Assembly of Virginia,* pp. 43–57.

72. *Ibid.*, Law of February 5, 1816.

Tenth Annual Report of the President and Directors of the Board of Public Works to the General Assembly of Virginia (Richmond, 1826), pp. 193–94. Subsequent citations of these reports will be to BPW with volume and year.

73. House of Delegates, *Journal* (1819–20), p. 144. See also Dunaway, *History of the James River and Kanawha Company,* Chap. III.

74. BPW IV (1820), Supplement, 3.

75. A fuller history of the Fund and Board is given in Goodrich, "The Virginia System of Mixed Enterprise." This includes tables of Current Revenues, Expenditures, and Investment, and of the Capital Account of the Fund, 1816–61.

76. BPW II (1818), 9. BPW XI (1827), 115. BPW XV (1831), 113.

77. The principle of deferred dividends on the state's stock had already been dropped, in spite of its ingenuity, and the Board was reorganized in 1831 to exclude the ten private citizens.

78. BPW XXIII (1829), 250, 254. Dunaway, *History of the James River and Kanawha Company,* Chaps. IV–VI.

79. Dunaway, *History of the James River and Kanawha Company,* pp. 187–88 and p. 118, note 2.

80. BPW XXXVI (1851), iii, iv. After this year the Board consisted of three commissioners, each elected to represent a geographic area, replacing the state officers who had served ex offico.

81. Turner, The Virginia Railroads, 1828–1860 (MS), pp. 7, 68, 73–74. BPW XXXVIII (1853–54), iv, vii, xvi–xvii.

82. BPW XL (1857), xii. BPW XXXIV (1849), 4.

83. BPW XX (1836), 13.

Letter Books of Board of Public Works, November 11, 1837 (MS in Virginia State Library).

An article attributed to Baldwin, "Board of Public Works in Virginia," in *North American Review*, VIII (1818), 1–26, pointed out that "the profession of a civil engineer [was] hardly known among us." His contributions have recently been discussed in Abbott, The Role of the Civil Engineer in Internal Improvements (MS).

84. BPW XIII (1829), 423; VII (1822), 66.

85. *Proceedings and Debates of the Constitutional Convention of 1829–1830*, p. 114. BPW XXVI (1841), 238–49, 251–53. The point is more fully developed in Ambler, *Sectionalism in Virginia from 1776 to 1861.*

86. BPW XIV (1830), 34; BPW XXII (1838), 5; BPW XXXVIII (1851), iv–xxii.

87. BPW XVII (1833), 44; BPW XXIII (1839), 76; BPW XXXI (1847), 16; BPW XXXIV (1849), 14. Dozier, *A History of the Atlantic Coast Line Railroad,* Chap. III. The original subscriptions to the Petersburg Railroad included something over $40,000 by the corporation of Petersburg and $160,000 by the state. A state loan was later converted into stock, and the stock turned over by the state in 1849 amounted to $325,500. Petersburg paid $10,000 for the City Point, and the state assets turned over consisted of $10,000 in stock and a $50,000 mortgage.

88. Heath, "Public Railroad Construction and the Development of Private Enterprise in the South Before 1861," p. 41, Table I-A. Heath, Public Co-operation (MS), pp. 125–31, gives greater detail.

89. Tables I and II in "The Virginia System" give the following figures:

Improvement Holdings, 1861	$39,343,808
Previous Disinvestment	2,074,718
Total Investment	$41,418,526

The "Report of the Select Committee on the subject of Internal Improvements," March 1860, House of Delegates (1859–60), *Doc. 86,* notes the following guarantees of company bonds:

Chesapeake and Ohio Canal Company	$500,000
Alexandria Canal Company	15,000
Richmond and Danville Railroad Company	200,000
Virginia Central Railroad Company	100,000
Total	$815,000

If this sum is added to that given above, the total is $42,333,526. The state also guaranteed municipal bonds for improvement purposes of $500,000 for Wheeling and $323,500 for Petersburg.

An indication of the distribution of this investment may be obtained from another tabulation made by the same committee:

Canal and Navigation Companies	$12,404,672
Railroads (including state railroads)	18,631,909
Turnpike Companies	2,305,176
Plank Road Companies	410,338
Bridge Companies	136,035
State Roads	1,765,544
Total	$35,653,674

This total may be brought into fairly close agreement with the figure given above for "Investment Holdings, 1861" by adding the substantial new investment made in 1860 and 1861 and an allowance for revaluation of the state's interest in the James River and Kanawha and by subtracting interest paid on certain loans and guaranteed bonds.

Heath, "Public Railroad Construction," Table I-A, gives the total of state investment in railroads through 1861 as $23,726,596.

90. "The Virginia System," Table I. "Report of the Select Committee," pp. 14–15. By contrast, the Board reported in 1851 that the state possessed 872 miles of "the most capacious and substantially constructed canals in the Union" and about 3,000 miles of turnpikes.

91. Sadove, *Transport Improvement and the Appalachian Barrier,* p. 269, speaks of the lost opportunity for "a second class transportation improvement" which might have joined the "Southeast with the Northwest at an earlier date, for example a better road between the James and the Kanawha rivers."

92. BPW XV (1831), 262.

93. *Annual Report of the Virginia Central Railroad Company,* 1850, quoted in Turner, *Chessie's Road,* p. 39.

94. "The Paint Rock line was completed to a connection with the Tennessee roads on January 28, 1882." Brown, *A State Movement in Railroad Development,* p. 229. This was a branch of the Western North Carolina. The connection with South Carolina was made later in the decade by the Spartanburg and Asheville. Both roads are now part of the Southern Railway System, and the "Carolina Special" provides daily through service between Cincinnati and Charleston. For the geographic setting, see Brown, pp. 2–5.

95. The phrase is from the Message of Governor George McDuffie,

November 1, 1836. South Carolina, *Legislative Proceedings,* 1836–37, p. 5.

96. Alfred G. Smith, Jr., *Economic Readjustment of an Old Cotton State,* Chap. V, contains an account of internal improvements in South Carolina of which I have made extensive use. This work is based on a doctoral dissertation accepted at Columbia University in 1954.

97. "Col. Senf's Account of the Santee Canal," ed. by Mabel L. Webber, *South Carolina Historical and Genealogical Magazine,* XXVIII (1927), p. 125. See also Porcher, *The History of the Santee Canal,* and Phillips, *A History of Transportation in the Eastern Cotton Belt to 1860,* pp. 35–43. The first president of the company was also governor of the state.

98. "Internal Improvements in South Carolina," attributed to John L. Sullivan, *North American Review,* XIII (1821), 149. The Engineer's Report for 1819 refers to a "distinguished citizen . . . who, without any emolument, devoted his whole time during the summer and fall months, to the direction of those works on Peedee River." Kohn (ed.), *Internal Improvements in South Carolina, 1817–1838,* p. 6 of Report. A gentleman who said that he was "far advanced in life and should be preparing to leave this world [but] still bless'd with health and vigor of spirits" offered his talents for work on the Catawba. Letter of John Barnett to V. A. Edwards, December 11, 1819 (MS in collection of South Carolina Archives Commission, Columbia, S.C.).

99. Report of the Committee on Internal Improvements, South Carolina Legislature, *Acts and Resolutions,* 1822, pp. 95–100. Message of Governor Robert Y. Hayne, *Legislative Proceedings,* 1833, p. 9. An act of 1838 provided for state subscriptions to turnpike companies to the amount of two-fifths of their capital stock. This provision remained almost completely unused but was sometimes cited in arguments for more systematic aid to railroads.

100. Derrick, *Centennial History of South Carolina Railroad,* pp. 59, 47.

Like the Baltimore and Ohio, the South Carolina experimented with sail power, and was reported to have discovered that a car could be made to "sail within four points of the wind." Charleston *Courier,* March 20, 1830, quoted by Derrick, p. 44. This book is the principal authority on the history of the railroad. See also Phillips, *A History of Transportation,* Chap. III.

101. Report of the Senate Committee on Finance, *Legislative Proceedings,* 1836, p. 62.

102. Charleston *Courier,* October 15, 1835, quoted in Derrick, *Centennial History of South Carolina,* p. 132.

103. The proponents recognized that the sale of cotton would require the building of factories in the Middle West, but Professor Smith has asked why it was supposed that the grain-growing Middle Westerners would begin eating rice. Smith, *Economic Adjustment of an Old Cotton State,* p. 163.

104. November 4, 1835, quoted in Derrick, *Centennial History of South Carolina,* pp. 135–36.

105. Baltimore *Patriot,* July 23, 1836, quoted in Phillips, *A History of Transportation,* p. 184.

106. The $2,000,000 par value of the South Carolina stock was counted in the assets of the L. C. & C., in spite of the fact that the transaction, in which the South Carolina stockholders were given a bonus of $20 per share, added $2,400,000 to the new company's liabilities. Moreover, the South Carolina stockholders were to receive cash payments for their old stock at a more rapid rate than that of the payments they would have to make on their new stock in the L. C. & C.

107. There has been much discussion of what the fate of the enterprise might have been if Calhoun had not withdrawn his support and if Hayne, the company's president, had not died in 1839: Van Deusen, *Economic Bases of Disunion in South Carolina,* pp. 228–37; Derrick, *Centennial History of South Carolina,* pp. 175–76; Jervey, *Robert Y. Hayne and His Times;* Cleveland, *Controversy between John C. Calhoun and Robert Y. Hayne as to the Proper Route of a Railroad from South Carolina to the West.* The immediate issue was over the proper route to the Tennessee Valley, but Jervey argued that Calhoun was also more impressed by the political advantages of a railroad connecting Southeast and Southwest via Memphis than by those of Hayne's connection between Southeast and Middle West.

108. The company also received $42,500 of state aid through absorption of the Pendleton Railroad and $5,000 from the town of Walhalla.

At the end of the fifties there was renewed agitation for a railroad by way of the French Broad on the part of citizens of Greenville and Spartanburg, South Carolina, and of Asheville and the counties of western North Carolina. Brown, *A State Movement in Railroad Developments,* p. 142. Since this was the route finally utilized a generation later, it might perhaps have been indicated on the map on p. 34 above; but it was the Blue Ridge project which attracted much greater attention and public aid in the fifties.

109. Smith, *Economic Adjustment of an Old Cotton State,* gives the total of initial state subscription, including that to the South Western Bank, as $2,651,600, which he compares with the total subscription reported in Heath, Public Co-operation in Railroad Construction (MS), of $4,242,600. Companies receiving railroad stock from the state in several cases had to dispose of it at moderate discounts. For the history of the various roads, see Phillips, *A History of Transportation,* pp. 335–55, 361–64.

110. Heath, "Public Railroad Construction," Table I-A, gives totals of $9,659,204 for state aid and $4,013,700 for municipal aid. There was no county aid.

111. Governor McDuffie, November 1, 1836, South Carolina *Legislative Proceedings,* 1836–37, p. 5.

112. In the thirties, when William Gregg, the industrialist, opposed the Charleston merchants' "unpractical projects of linking their port with the Ohio Valley by railroads built with state subsidies," he was in a small minority, at least of vocal opinion. In the late fifties, his reasoned opposition had much to do with holding down the size of the state's subscription to the Blue Ridge. Derrick, *Centennial History of South Carolina,* p. 176. Van Deusen, *Economic Bases of Disunion in South Carolina,* pp. 251–53.

113. "North Carolina evidently holds, unrivalled, the best ground of communication with much of the west and all the south west." Board for Internal Improvements, *Annual Report,* 1833, p. 15. Brown, *A State Movement in Railroad Development,* organizes its argument largely around the east-west, north-south competition. This book is the principal authority on North Carolina improvements. The earlier work is Weaver, "Internal Improvements in North Carolina previous to 1860."

114. Archibald D. Murphey, *Papers,* ed. by William Henry Hoyt (Raleigh, North Carolina Historical Commission [19–?]), II, 150. See also pp. 19–29, 83–87, 96–195.

115. Board for Internal Improvements, *Annual Report,* 1821, p. xix.

116. Murphey, *Papers,* II, 24, 44, 23.

117. Board for Internal Improvements, *Annual Report,* 1821, p. xxv. *Ibid.,* 1825, p. 3. *Ibid.,* 1833, cited in Murphey, *Papers,* p. 102. State subscriptions to the Cape Fear and Deep River Navigation Company and to a new Neuse River company were made in 1849 and 1850. Weaver, "Internal Improvements in North Carolina," pp. 55, 65.

118. State bonds for this road were sold at discounts of about

$112,000 and Craven County bonds at a discount of about $6,000. Brown, *A State Movement in Railroad Development,* p. 222. The remainder of Chap. VII gives a full account.

119. Asheville *News,* April 8, 1858, quoted in Brown, *A State Movement in Railroad Development,* p. 139.

120. Heath's figure for state aid, including endorsements, is $10,-655,979; for municipal aid, mainly from Wilmington, $427,000; and for county aid, mainly "from the land-locked and long-neglected western counties," $850,000. "Public Railroad Construction," Table I-A. Public Cooperation in Railroad Construction (MS), pp. 108, 237.

121. Internal Improvements Convention, Raleigh, July 4, 1833, quoted in Dozier, *A History of the Atlantic Coast Line Railroad,* pp. 54–55; pp. 56–66 are devoted to the Wilmington and Weldon.

122. Brown, *A State Movement in Railroad Development,* p. 280. Chap. XV is devoted to an appraisal of the "trunk line idea." The author suggests (pp. 282–83) that a trunk line through North Carolina, but based on Norfolk rather than a North Carolina port, might have been "not altogether incomparable with the Chesapeake and Ohio or the Norfolk and Western."

123. *Letters and Correspondence of John C. Calhoun,* ed. by J. F. Jameson (Washington: American Historical Association, 1900), p. 412.

124. Gallatin's Report, p. 732. This would, he said, serve to "discharge the Mississippi into the Atlantic."

125. The reader interested in internal improvements is particularly fortunate in the case of ante-bellum Georgia. He can use the earlier work of Phillips, *A History of Transportation,* of which Chap. VII points out the uniqueness of the Western and Atlantic case; and the dean of modern students of internal improvements, Professor Milton S. Heath, has covered the subject fully in *Constructive Liberalism,* esp. Chaps. X and XI.

126. Heath, *Constructive Liberalism,* p. 245.

127. *Ibid.,* p. 266.

128. Phillips, *A History of Transportation,* pp. 307–10, lays particular stress on the influence of the officials of the three railroad companies, pointing out that the bill was sponsored in the legislature by a member who was also president of the Central of Georgia. Heath gives greater emphasis to the role of Governor Lumpkin. At a time and in a region when railway-building was thought of as having so largely a civic character, the difference is hardly decisive. Lumpkin himself was a member of the first board of directors of the Georgia Railroad.

129. Phillips, *A History of Transportation*, p. 308.

130. Heath, *Constructive Liberalism*, pp. 368–74 including Tables 25 and 26.

131. "The city of Savannah supplied some 80 per cent of the capital of the Savannah, Albany and Gulf Railroad, yet private shareholders with the remaining 20 per cent held the entire voting power in the company." *Ibid.*, p. 435, n. 80.

4. FROM BOSTON TO MOBILE

1. Governor King's message of December 30, 1850, quoted in Primm, *Economic Policy in the Development of a Western State*, p. 101.

2. J. Thomas Scharf, *History of Delaware*, 1609–1888 (Philadelphia: L. J. Richards, 1888), Vol. I, Chap. XXII, esp. p. 423.

3. Delaware, *Laws*, VIII, 233–36, January 22, 1833; X, 241–42, February 11, 1839; IX, 17–26, June 20, 1836; XI, 293–94, December 28, 1855; 723–24, February 24, 1859.

The city of Wilmington did not make the subscription permitted by the single act passed in this period authorizing municipal aid to a railroad. Goodrich, "Local Government Planning," Table I. Delaware *Laws*, XI, 311, March 2, 1855; XII, 136–43, March 5, 1861.

4. Cranmer, The New Jersey Canals (MS), esp. p. 3.

5. *Ibid.*, p. 331, Table XVI. See also Burgess and Kennedy, *Centennial History of the Pennsylvania Railroad Company*, pp. 233–70.

6. Quoted in Cranmer, The New Jersey Canals (MS), p. 114.

7. "As anticipated, the developmental Morris accomplished a renaissance of the iron industry and aided materially in the industrialization of North Jersey." *Ibid.*, p. 349. Detailed evidence supporting this conclusion, as well as tables of tonnage and operating results, are given by the author.

8. Stock of the company which brought 120 in 1825, was down to 20 in 1826. Cranmer, The New Jersey Canals (MS), on p. 131, cites the following description from a contemporary newspaper of the bargain-hunters of 1824: "The wall street bees, attracted hither by the savor of New Jersey honey, are buzzing about the State House and looking out sharply for the delicious flowers of good bargains on which to feed and fatten."

9. *Ibid.*, p. 349. Note also the title of George A. Reilly, The Camden and Amboy Railroad in New Jersey Politics (Unpublished dissertation, Columbia University, 1951).

10. It is for this reason that New England is treated here, rather than in the chapter on "The Appalachian Competition," in spite of the fact that its mountains are part of the Appalachian system. But Rubin, Imitation by Canal or Innovation by Railroad (MS), stresses the similarity between the Berkshires as an obstacle to Boston's trade with the West and the Alleghenies as the obstacle to the trade of Philadelphia and Baltimore.

11. Kirkland, *Men, Cities and Transportation,* I, 232. This important work is the principal authority on New England improvements. For the Middlesex Canal, see also pp. 20–21 above.

12. *Ibid.,* p. 124.

13. *Speech of Otis Clapp, of Boston, on the Bill for Loaning the Credit of the State . . . in the Legislature of Massachusetts, May 8th, 1851,* (pamphlet in Columbia University Libraries), p. 8.

14. *Speech of T. G. Cary on the Use of the Credit of the State for the Hoosac Tunnel in the Senate of Massachusetts, May 18, 1853* (Boston: J. M. Hewes, 1853), p. 34. Cary was an opposition spokesman.

15. F. W. Bird, "The Road to Ruin: or, the Decline and Fall of the Hoosac Tunnel" (Boston: Wright & Potter, 1862), p. 24. The total of town subscriptions was $145,600. Kirkland, *Men, Cities and Transportation,* I, 403, 408.

16. Kistler, *The Rise of Railroads in the Connecticut River Valley,* pp. 155–56.

17. New Hampshire repealed in 1840 its one act authorizing a town subscription before such a subscription was made. Kirkland, *Men, Cities and Transportation,* I, 326. See also Goodrich, "Local Government Planning," Table I.

18. Kirkland, *Men, Cities and Transportation,* I, 180.

19. Poor, *The First International Railway,* p. 51. Laura Poor was J. A. Poor's daughter. A great grand-nephew gives a briefer and less laudatory account. Chandler, *Henry Varnum Poor,* pp. 11–19.

20. Kirkland, *Men, Cities and Transportation,* I, 208.

21. Provincial and Montreal city bonds, amounting to over £500,000, made up more than two thirds of the capital of the St. Lawrence and Atlantic. The Grand Trunk system as a whole received much greater public aid. Glazebrook, *A History of Transportation in Canada,* pp. 164, 174–75.

22. Portland, January 5, 1858. Quoted in Kirkland, *Men, Cities and Transportation,* I, 213–14.

23. Poor, *The First International Railway,* p. 154.

24. Bogart, *Internal Improvements and State Debt in Ohio,* p. 54. This is the standard work on the subject. An important recent contribution is made by Segal, Canal Cycles (MS), Chap. IV.

25. See above, pp. 39–40. Five hundred thousand acres were also given to the state of Ohio for general canal purposes, but the proceeds of their sale, amounting to more than $600,000, were applied to payment of interest on the public debt. Bogart, *Internal Improvements and State Debt,* p. 37. "Between 1833 and 1835, Miami land sale receipts totalled $304,895." Segal, Canal Cycles (MS), p. 227, note 18.

26. Black, *The Little Miami Railroad,* pp. 44–45.

27. Segal, Canal Cycles (MS), p. 219. This study gives a full account of the Commissioners' financial expedients.

28. *Ibid.,* Appendix Table 3, pp. 279–80, and pp. 229–30. The quotation is from the act of March 14, 1836, relating to the Walhonding Canal. The Board of Public Works was required to make this finding before work could be started. Similar provisions appeared in the measures authorizing the Hocking Canal and the Muskingum Improvement. See also Bogart, *Internal Improvements and State Debt,* pp. 76–88; Taylor, *The Transportation Revolution, 1815–1860,* pp. 45–47.

29. This figure is based on the unpublished research of Harvey H. Segal. No systematic study of local aid in Ohio has appeared, but a number of illustrations are given in Black, *The Little Miami Railroad.* On p. 117, he interprets Cincinnati's purchase of waterfront property from the Ohio and Mississippi for $500,000 in 1855 as a subterfuge for aid that could no longer be extended directly.

30. Bogart, *Internal Improvements and State Debt,* p. 118. On p. 53, he notes the suspicious circumstances under which the bankrupt Pennsylvania and Ohio Canal Company, "to which the state had contributed about $450,000," was transferred for nothing to a railroad owned by the governor of the state; and on p. 110 he suggests the possibility of corruption in the lease of the canals. It should, however, be noted that, at the time of the lease, the canals had shown an operating deficit for the preceding three years.

31. Benton, "The Wabash Trade Route in the Development of the Old Northwest," pp. 33, 94. For the financial results of the canal, see p. 88, note 64, and for economic results, Chap. III.

32. Daniels, "The Village at the End of the Road," esp. pp. 17–19, 29–31, 82–85.

33. On the basis of a number of newspaper files and county histories. A full study of local aid in Indiana has not been made. The number of

special laws authorizing it was as follows: in the *1830s*, 13; in the *1840s*, 23; in the *1850s*, 9.

34. Daniels, "The Village at the End of the Road," p. 81, note 66, quoting Governor Ray.

35. Putnam, *The Illinois and Michigan Canal.* This is the standard work on the subject. The indebtedness was discharged and the canal turned back to the state in 1871. *Ibid.*, pp. 64–65. Chicago, which had about 1800 inhabitants in 1830, grew from a population of 20,000 to nearly 75,000 in the six years from the opening of the canal to the beginning of railroad competition. *Ibid.*, p. 107. See also Pease, *The Frontier State, 1818–1848,* Chaps. X, XI, XVII.

36. Pease, *The Frontier State, 1818–1848.*

Gates, *The Illinois Central Railroad, and Its Colonization Work,* pp. 22–23.

Krenkel, *Illinois Internal Improvements, 1818–1848.*

Chapter X of Pease describes the process of logrolling and relates it to the contest over the choice of a state capital. Krenkel's work, which is based on his 1937 doctoral dissertation at the University of Illinois, lays less stress on logrolling (pp. 72–74, 84–88) and more on the general enthusiasm for improvements. He remarks (p. 59) that "in boom times like those of 1836 no section would sit back to see its neighbor gain an advantage by acquiring a much needed means of communication."

37. Krenkel, *Illinois Internal Improvements,* p. 204, says "approximately one fortieth." The state's entire railroad assets brought less than a $1,000,000, mainly in depreciated state obligations. *Ibid.*, p. 200.

38. Gates, *The Illinois Central Railroad,* pp. 83, 139.

Jacklin, Local Aid to Railroads in Illinois, 1848–1870 (MS), esp. Appendix Chart A.

39. See below, pp. 169–72.

40. Kent Sagendorph, *Stevens Thomson Mason: Misunderstood Patriot* (New York: Dutton, 1947), p. 338. This book tells the story from the point of view of the Democratic governor who was the principal figure in the Michigan program. Abandonment and sale were the work of his Whig successors. See also Pearson, *An American Railroad Builder: John Murray Forbes,* and MacGill and others, *History of Transportation in the United States before 1860,* pp. 503–05.

41. John W. Brooks, in a letter to Erastus Corning, April 5, 1853, quoted in Neu, "The Building of the Sault Canal: 1852–1855," p. 35.

42. Peter White in Charles Moore, ed., *The Saint Mary's Falls Canal* (Detroit: Semi-centennial Commission, 1907), p. 30.

43. *House Journal,* 1842–43, 1st session, Governor's Message.

44. Merk, *Economic History of Wisconsin during the Civil War Decade,* pp. 279–284.

45. *The Banker's Magazine,* 1859, p. 717, cited in Phelan, "The Financial History of Wisconsin," p. 232. There were also local government guarantees of railroad bonds, and their total to 1858 is given as $1,614,000 in Larson, "A Financial and Administrative History of Milwaukee," pp. 74–75. See also Meyer, "A History of Early Railroad Legislation in Wisconsin," pp. 206–300.

46. ARJ XXX (1857), 379.

47. Merk, *Economic History of Wisconsin during the Civil War Decade,* Chap. IX, describes the indignation of Wisconsin farmers who had not only bonded their communities but also mortgaged their own farms for railroad subscriptions.

48. Beard, "Local Aid to Railroads in Iowa." No quantitative study has been made. Beard cites guesses ranging from six to eleven million for the total of local aid. The $7,000,000 figure is used in Cleveland and Powell, *Railroad Promotion and Capitalization in the United States,* p. 205.

49. Internal improvements in Missouri have received unusually full treatment. The early work is Million, *State Aid to Railways in Missouri;* Lopata, *Local Aid to Railways in Missouri,* is the first published work containing a thoroughly documented study of local aid in any state; Primm, *Economic Policy in the Development of a Western State,* Chaps. V and VI, adds interpretation of the attitudes of proponents and opponents.

See also Belcher, *The Economic Rivalry between St. Louis and Chicago, 1850–1880,* Chap. V, "St. Louis and the Railroads," Riegel, *The Story of the Western Railroads,* Chaps. II and IV; Russel, *Improvement of Communication with the Pacific Coast,* Chap. VIII.

The quotation is from St. Louis *Missouri Argus,* March 11, 1836. The editor expressed the hope that the program would be "as broad as the wants of the common wealth." Cited in Primm, *Economic Policy in the Development of a Western State,* pp. 79–80.

50. But note the conclusion of Belcher, *The Economic Rivalry between St. Louis and Chicago, 1850–1880,* that St. Louis did not sufficiently realize the value of railroads.

51. Letter of March 23, 1853, from Secretary of the North Missouri Railroad Company, quoted in Lopata, *Local Aid to Railways in Missouri,* p. 32, note 13. Lopata (p. 32) gives the following interpretation

of the policy: "What we propose to do is to lend the credit of the company to the *taxpayers,* and incidentally to the railroads, to purchase a block of stock en masse."

52. Million, *State Aid to Railways in Missouri,* p. 121, estimates the total expended on the construction of the Missouri roads to 1860 as about $40,000,000. The accompanying table shows the distribution of state and local aid to the Missouri railroads during the decade 1850–60. The amounts of aid are taken from Lopata, *Local Aid to Railways in Missouri,* pp. 16–19, 59, and the mileage from Belcher, *The Economic Rivalry between St. Louis and Chicago, 1850–1880,* p. 91.

STATE AND LOCAL AID, 1850–60 *

Railroad	*State Loans*	*Local Government Subscriptions*	*Total*	MILEAGE, 1860
Hannibal & St. Joseph	$3,000,000	$ 202,500	$3,202,500	206
Pacific	7,000,000	2,859,100	9,859,100	189
North Missouri	4,350,000	2,217,883	6,567,883	168
St. Louis & Iron Mountain	3,501,000	1,705,000	5,206,000	86
Southwest Branch	4,500,000	63,333	4,563,333	77
Platte County	700,000	187,500	887,500	44
Cairo and Fulton	650,000	400,000	1,050,000	26
Ohio and Mississippi	—	700,000	700,000	†
Total	$23,701,000	$8,335,316	$32,036,316	796

* The aid figures given above exclude $1,249,000 authorized but not extended by the state and $784,184 authorized but not extended by local authorities. No allowance is made for the discounts taken when the railroads sold the state and local bonds. These were substantial after 1857.

† No mileage in Missouri.

53. Law of December 20 and Memorial of December 21, 1805, cited in Gallatin Report, p. 822. It will be recalled that this improvement formed part of Gallatin's plan.

54. Trescott, "The Louisville and Portland Canal Company, 1825–1874," esp. pp. 689–690, which give quotations from Kentucky *House Journal,* 36th Session (1827–28), p. 18, and *Congressional Globe,* 23d Congress, 2d Session (1835), p. 202. Also see below, p. 169.

After the company had become profitable, the state rejected a proposal to join with the city of Louisville and the federal government in buying out the privately held stock in order to convert the canal into a toll-free improvement. For an abstract of Kentucky and United States legislation from 1825 to 1857, see "Report of Committee on the Best Mode of Improving Navigation at the Falls of the Ohio" (Cincinnati, 1859), pp. 35–59.

55. ARJ XXX (1857), 488.

56. State, county and municipal totals, as for other southern states, are taken from Heath, "Public Railroad Construction," Table I-A.

57. Cincinnati *Gazette,* 1851, cited in Daniels, *The Village at the End of the Road,* p. 67. Pages 65–78 describe the competition of the three river cities of Cincinnati, Louisville, and Madison, Indiana, for the trade of central Indiana.

58. Clark, *The Beginning of the L. & N.*

59. Folmsbee, *Sectionalism and Internal Improvements in Tennessee.*

60. *Ibid.,* p. 63.

61. *Ibid.,* p. 106, quoting "A Citizen" in Knoxville *Register,* June 22, 1831.

62. *Ibid.,* p. 266, quoting the October, 1845, message of Governor James G. Jones. The italics are the Governor's.

63. Heath, Public Cooperation in Railroad Construction (MS), pp. 213–26.

64. Martin, "Internal Improvements in Alabama," Chap. III. Page 76 gives quotations from the veto messages of Governor John A. Winston. For federal aid, see below, p. 170.

65. Peter J. Hamilton, *Mobile of the Five Flags* (Mobile: Gill Printing Co., 1913), p. 248.

66. DeBow, *Industrial Resources, etc. of the Southern and Western,* II, 439. See also *Hunt's Merchant Magazine,* Vol. 19 (1848), pp. 579–93, for an enthusiastic endorsement of the project.

67. This modified an earlier law of 1850. DeBow, *Industrial Resources, etc. of the Southern and Western,* II, 535. Charles Grayson Summersell, *Mobile; History of a Seaport Town* (University, Ala.: University of Alabama Press, 1949), p. 24.

68. DeBow, *Industrial Resources, etc. of the Southern and Western,* II, 484.

69. Cairns, The Response of New Orleans to the Diversion of Trade from the Mississippi River, 1845–1860 (MS).

70. ARJ XXV (1852), 488; XXVI (1853), 34.

71. DeBow, *Industrial Resources, etc. of the Southern and Western,* II, 484.

72. Cotterill, "Southern Railroads, 1850–1860," pp. 398–405; Heath, Public Cooperation in Railroad Construction (MS), pp. 166–77.

73. The point is made in Heath, Public Cooperation in Railroad Construction (MS), pp. 191–200. See also Cairns, The Response of

New Orleans to the Diversion of Trade from the Mississippi River (MS), and Corliss, *Mail Line of Mid-America,* Chaps. 17, 18 and 24, esp. p. 188.

74. Heath, Public Cooperation in Railroad Construction (MS), pp. 155–63.

75. Reed, *A History of the Texas Railroads,* p. 81.

76. *Ibid.,* p. 48, quoting John W. Dancy who was a Senator under the Republic and State.

77. Governor Wickliffe of Louisiana, quoted in ARJ XXX (1857), 84.

78. Reed, *A History of the Texas Railroads,* Chaps. 6–21, 24. Tabulations of the amount of aid are given on pp. 134, 137, and 145.

79. Adams, *Public Debts,* p. 339.

80. See above, pp. 34–35, map and accompanying table, and Goodrich, "The Gallatin Plan after One Hundred and Fifty Years." The Louisville and Portland Canal, which was included in the Gallatin Plan but is not shown on the map, was a federal-private enterprise.

81. See above, p. 17.

5. AN ERA OF NATIONAL SUBSIDY

1. *Congressional Globe,* 30th Congress, 1st Session, p. 535.

2. Trescott, "The Louisville and Portland Canal Company, 1825–1874."

3. These measures are examined in detail in Haney, *A Congressional History of Railways . . . to 1850,* Chaps. XI–XIX. For a tabulation of grants under the 1841 act, see Donaldson, *The Public Domain,* p. 255.

4. *Congressional Globe,* 31st Congress, 1st Session, pp. 845, 848, 1437, 869, 851. Donaldson, *The Public Domain,* pp. 262–64, contains a statement by J. M. Cutts purporting to embody Senator Douglas's account of his negotiations in the South after the entire delegations from Mississippi and Alabama had voted against the Illinois Central measure and of his subsequent strategy in the Senate.

5. *Ibid.,* pp. 536, 849, 903. The bill passed the Senate, 26 to 14, and the House, 101 to 73. *Ibid.,* pp. 904, 1838. Gates, *The Illinois Central Railroad,* pp. 27–43, analyzes the political alignments in Congress, suggesting that a part of the bill's support might be explained in terms of interests of eastern investors in Illinois lands and Illinois improvement bonds.

6. ARJ XXVI (1953), 470. The colonization efforts of the Illinois

Central form the principal theme of Gates, *The Illinois Central Railroad.*

7. The total actually patented under these measures, after a number of reassignments, was about 18,500,000 acres. Donaldson, *The Public Domain,* pp. 269–73. See also Haney, *A Congressional History of Railways . . . from 1850 to 1887,* pp. 17–18, 185.

8. Russel, *Improvement of Communication with the Pacific Coast as an Issue in American Politics, 1783–1864.*

9. Jackson, *Wagon Roads West.*

10. The Tehuantepec Company received a mail contract for one year at what was clearly a subsidy rate, and the Panama Company asked but failed to receive a twenty-year contract for the carriage of government freight and personnel. Russel, *Improvement of Communication with the Pacific Coast,* pp. 238–39, 59–61.

11. F. W. Lander, "Report of a Reconnaissance from Puget Sound," in "Reports of Explorations and Surveys to Ascertain the Most Practicable and Economic Route for a Railroad from the Mississippi River to the Pacific Ocean," II, 8–24. These Reports will be cited as Pacific Railway Surveys.

12. *Congressional Globe,* Thirtieth Congress, Second Session, p. 472. The various proposals are analyzed in Haney, *A Congressional History of Railways . . . to 1850,* Book III, Part II; in *A Congressional History of Railways . . . from 1850 to 1887,* Book II; and, with greater attention to political and sectional controversy, in Russel, *Improvement of Communication with the Pacific Coast.*

13. ARJ XXXVIII (1865), 345, 891–92. See also *Congressional Globe,* 37th Congress, 3d Session, pp. 1180, 1240–41.

14. Chandler, *Henry Varnum Poor,* pp. 59–60.

15. Message to Congress, December 6, 1858.

16. ARJ XXX (1857), 25.

17. May 26, 1847, cited in *House Report No. 140,* 31st Congress, 1st Session, p. 52.

18. *Ibid.,* esp. Appendix, p. 79 (Whitney letter of February 24, 1849) and pp. 73–74 (resolution, Cincinnati Chamber of Commerce, October 12, 1849).

19. *Hunt's Merchant Magazine,* XXIII, 124, cited in Haney, *A Congressional History of Railways . . . to 1850,* p. 413, note 14.

20. *House Report No. 140,* 31st Congress, 1st Session, p. 3.

21. *Grant County Herald,* cited in Meyer, *A History of Early Railroad Legislation in Wisconsin,* p. 255. As an illustration of the ambi-

tions of a Wisconsin village, the same author (p. 258) cites the proposed Port Ulao, Lake Superior, Lake of the Woods, & Behring Straits Transit Company.

22. *Congressional Globe,* 32d Congress, 2d Session, pp. 799, 815, 996, cited in Haney, *A Congressional History of Railroads . . . 1850–1887,* pp. 54–55.

23. Pacific Railway Surveys, Vol. I, "Report of the Secretary of War," esp. pp. 11, 20–21, 29–30. The following table, adapted and simplified from p. 31, summarizes the main points in the comparison:

Route	*Estimated Mileage*	*"Equated" Mileage*	*Highest Pass (in feet)*	*Estimated Cost ($000)*
47th–49th Parallels St. Paul-Vancouver	1,864	2,207	6,044	130,781
41st–42d Parallels Council Bluffs, Ia., to Benicia, Cal.	2,032	2,583	8,373	116,095
38th–39th Parallels Westport, Mo., to San Francisco	2,080	3,125	10,032	*
35th Parallel Fort Smith, Ark., to San Pedro	1,892	2,816	7,472	169,210
32d Parallel Fulton, Ark., to San Pedro	1,618	2,239	5,717	68,970

* No estimate given as route was considered impracticable.

"Equated mileage" represented an attempt to allow for differences in the sum of ascents and descents by the various routes. Cost figures were thought to be more significant in comparing the different routes than as estimates of the absolute cost by any route. That for the 35th-parallel route was regarded as much too high, and a revision by the Topographical Engineers in 1856 reduced the figure to $92,000,000. *Ibid.,* Vol. VII, "Conclusion," no. 5, p. 37.

24. "Nature has clearly indicated the northern pathway for the commerce from the future mart of Asiatic trade to this country and Europe. The great lakes carry us water-borne half-way across the continent." *Ibid.,* Vol. I, "Report by I. I. Stevens," p. 114. See also pp. 33, 52.

25. These two points are made by Russel, *Improvement of Communication with the Pacific Coast,* pp. 180, 182–83, giving actual mileages, as against the estimated mileages cited in footnote 23 above, of 1,814 for the northern, of 1,775 for the central route, and of 1,722 and 1,720

for the 35th and 32d parallels respectively. See also Albright, *Official Explorations for Pacific Railroads, 1853–1855,* which declares (p. 158) that "there can be no doubt that [Davis] was absolutely unpartisan" in his judgment in favor of the 32nd parallel route.

26. Russel, *Improvement of Communication with the Pacific Coast,* Chap. IX.

27. E. Haskett Derby, in *Hunt's Merchants Magazine,* XXXV, 659–80, cited in *ibid.,* p. 183.

28. Note for example Russel's statement (*ibid.,* p. 195) referring to the 1854–55 session: "There were about one hundred thirty members in the House who wanted to extend federal aid to a Pacific railroad; only seventy-five or eighty who were opposed. Yet no bill was finally adopted."

29. *Congressional Globe,* 37th Congress, 2d Session, pp. 2808, 1706.

30. Russel, *Improvement of Communication with the Pacific Coast,* Chaps. XIX, XX, esp. p. 316. On p. 302, the author describes the provision fixing the initial point of the main line on the 100th meridian in Nebraska near the Kansas line as "a fair arrangement between the two great conflicting interests." See also *Congressional Globe,* 37th Congress, 2d Session, p. 2750.

31. Senator Howe of Wisconsin, *Congressional Globe,* 37th Congress, 2d Session, p. 2810. Senator Sherman of Ohio declared (*ibid.,* p. 2784) that "there is no propriety in the Government doing anything more than to leap over the desert." For threats to vote against the 1864 measure if it were loaded with one more branch, see *Congressional Globe,* 38th Congress, 1st Session, p. 2421.

32. *Congressional Globe,* 38th Congress, 1st Session, pp. 3149, 3182, 3244, 2652. Pruyn's amendment providing for government construction was defeated by a vote of 72 to 20. Note also the following statement made in the 1862 debates: "There is no corporation or body of men or interest which will furnish any considerable proportion of the outlay. We deceive ourselves if we think to get any amount worth reckoning out of private individuals." *Congressional Globe,* 37th Congress, 2nd Session, p. 1707.

33. *Congressional Globe,* 38th Congress, 1st Session, pp. 3152, 3150, 2355.

34. Reports of two House of Representatives committees were published in February, 1873: 42d Congress, 3d Session, *House Report No. 78,* "Affairs of the Union Pacific Railroad Company," and *House Re-*

port No. 77, "Credit Mobilier Investigation." These will be cited, from the names of the committee chairmen, as the Wilson Report and Poland Report respectively.

Fogel, The Union Pacific Railroad (MS), gives the following table of bond issues through 1872:

	Face Value	*Proceeds*
Government Bonds	$27,236,512	$27,145,163
First Mortgage Bonds	27,213,000	23,718,009
Land-Grant Bonds	10,400,000	6,063,992
Income Bonds	9,335,000	6,536,600

Of the total cost given for construction, nearly $8,500,000 was paid directly by the Union Pacific and $51,000,000 through the Credit Mobilier arrangements. The former figure includes engineering fees, construction of fences, stations and other buildings, and some equipment, as well as the costs of short-term borrowing. Since the Union Pacific sold the section of its line from Ogden to Promontory Point, Utah, to the Central Pacific shortly after the roads were completed for a little more than $2,000,000, the net cost of construction might be regarded as about $57,000,000. Disregard of the direct payments by the Union Pacific, as Fogel points out, led the Wilson Committee to the erroneous and often repeated statement that the first mortgage and government bonds were more than sufficient to pay for the cost of the road. Trottman, *History of the Union Pacific,* Chaps. II–IV, refers to the use of the income bonds for the purchase of equipment and rolling stock.

35. The Wilson Report, pp. xiv–xvii, estimated the current value of the profits as $23,366,319, valuing Union Pacific first mortgage bonds at from 80 to 90 and its stock at 30. The Committee's figure has been commonly used in later accounts. Donely, The Construction of the Union Pacific Railroad by the Credit Mobilier of America (MS), and Fogel, The Union Pacific Railroad (MS), have demonstrated that the Credit Mobilier could not have received or retained enough bonds to justify the Committee estimate.

By an examination of the receipts and disbursements of the Credit Mobilier under the various contracts, Donely estimates the current value of its profit as $13,691,435. This implies that the securities received by the construction company consisted almost entirely of common stock.

Fogel sets an upper limit of about $16,500,000 to the possible profits of the Credit Mobilier in current terms on the assumptions (*a*) that it took all the cash of the Union Pacific not known to have been disposed of otherwise, (*b*) that the Credit Mobilier had no operating expenses of its own, and (*c*) that the market value of Union Pacific stock was 30. The operating expenses of the Credit Mobilier could not have been large, since the Union Pacific itself supervised the work of the subcontractors. Allowing for expenses and for the possibility that the market value of the stock was less than 30, Fogel sets $13,000,000 as the lower limit of Credit Mobilier profits in current terms. His study makes full use of the reports prepared by accountants for the 1887 "Report of the Commission and of the Minority Commissioner of the United States Pacific Railway Commission."

Cash subscriptions to the Credit Mobilier amounted to a total of $3,750,000, part of which was used to buy out the stock of the early subscribers to the Union Pacific. Subscribers to the $1,250,000 of Credit Mobilier stock sold in 1867 and 1868 were also to pay in $225,000 as a partial offset to a dividend of $5,000,000 in Union Pacific stock; but apparently the actual payment was only $171,000 on $3,800,000 of stock. These two sums represent the entire equity investment until after the completion of the road, although Durant bore an additional share of the risk by advancing $435,000 in the early years. In 1870, after the road had been completed, the stockholders of the Credit Mobilier paid $4,672,000 for $5,840,000 of Union Pacific income bonds and $11,680,000 of Union Pacific stock. If, as the Wilson Committee believed, the income bonds were worth 80 at the time, the stock was in effect a stock dividend. If the bonds were worth less, a part of the purchase was equity capital. The profits of the Credit Mobilier stockholders on this transaction are included in Fogel's upper-limit figure. Wilson Report, pp. xvi, 630, 645.

Since the subscribers to the second issue of Credit Mobilier stock also received a bonus of $1,250,000 in Union Pacific first mortgage bonds, they obtained profits at a higher rate than those who invested the original $2,500,000. The shares distributed in Congress belonged to this quicker-profit category.

For the rate of turnover of the working capital during construction, see Wilson Report, pp. 337, 494, 563.

36. Charles Francis Adams and Horace F. Clark, each of whom was president of the Union Pacific Railroad after the period under discussion. Wilson Report, p. 987; Poland Report, p. 431.

37. Wilson Report, p. 162. The sentence ends as follows: "the resources furnished by the government being more than ample for the

purpose." This illustrates the characteristic disregard of the contribution of the first mortgage bonds.

38. Poland Report, p. 19.

39. The Fogel essay emphasizes the first of these factors and the Donely essay the second. Both problems were solved before the end of 1867.

40. Wilson Report, p. 568. See also Poland Report, p. 94. A government director of the Union Pacific, whose term expired in 1867, reported for his period that: "The sub-contracts were let with a good deal of care and a good deal of cutting down of prices, so that the sub-contractors made but a fair profit." Wilson Report, p. 653. The Poland Committee recommended expulsion of Ames and another Congressman. The House decided merely to reprimand them. For "The Fisk 'Raid'," see Trottman, *History of the Union Pacific,* pp. 47–50.

41. The 1864 act limited to 150 miles the distance to which the Central Pacific could build east of the California line, but the original provision was restored in 1866. Collis P. Huntington of the Central Pacific is said to have made the following reference to the 150-mile clause: "I said to Mr. Union Pacific when I saw it, I would take that out as soon as I wanted out. In 1866 I went to Washington." Daggett, *Chapters on the History of the Southern Pacific,* p. 50.

42. *Ibid.,* Chaps. I–IV. The author believes that the Central Pacific failed to find the best route across the Sierras, since the pass later used by the Western Pacific had easier grades and was 2,000 feet lower. *Ibid.,* pp. 66–67. The amount of subsidy was somewhat increased by a ruling which established the western base of the Sierras at a point at which little slope was visible to the naked eye. *Ibid.,* pp. 59–60. Page 21 cites Leland Stanford's figures of the receipts of the Central Pacific and Western Pacific railroads to December 31, 1869. The table makes no direct reference to stock subscriptions.

43. *Ibid.,* p. 81. See also *ibid.,* Chap. XII, "The Southern Pacific and Politics." Illustrations from the actions of the Central Pacific–Southern Pacific group are cited by the United States Pacific Railway Commission in support of the following statement: "It is the judgment of the Commission that moneys of all the bond-aided roads have been used for the purpose of influencing legislation. There is no direct proof . . . that actual bribery was resorted to. But it is impossible to read the evidence . . . without reaching the conclusion that very large sums of money have been used improperly in connection with legislation." *Report . . . of the Pacific Railway Commission,* p. 121.

44. Poland Report, p. 37; Wilson Report, pp. 431, 432.

45. "Reports of the commissioners who examined the Central Pacific and Union Pacific Railroads in 1869 and 1874," in 44th Congress, 1st Session, *House Report No. 440,* pp. 132ff., esp. pp. 136, 140.

46. For the reports from 1864 to 1885, see 49th Congress, 1st Session, *Senate Executive Documents No. 69,* pp. 109ff., and 44th Congress, 1st Session, *House Executive Documents, No. 180.*

47. Poland Report, pp. x–xiv, xix. The director was James S. Brooks, who was a Congressman at the time of the investigation and one of the two censured by the House.

48. Jesse L. Williams. The advantages of the route are stated in his reports of November 23 and December 1, 1866. Fogel, The Union Pacific Railroad (MS), says that Williams was the first man to write to the Secretary of the Treasury protesting against the profits of the Credit Mobilier.

49. Wilson Report, pp. 652–54, 67, 232, 367.

50. *Ibid.,* pp. 227–30, 535, 495, 433.

51. *Report of . . . the Pacific Railway Commission,* pp. 152, 19, 34.

52. Russel, *Improvement of Communication with the Pacific Coast,* pp. 320–22; Haney, *A Congressional History of Railways . . . from 1850 to 1887,* Chaps. IX, X, XII.

53. In claiming that the construction of the Union Pacific had "resulted in the solution of the transcontinental railroad problem," Charles Francis Adams declared: "The moment it was found that such a road could be built and operated at a profit other roads were built." Wilson Report, p. 982. This assertion seems hardly borne out by the times at which or the terms under which the other railroads were constructed.

54. Riegel, *The Story of the Western Railroads,* Chaps. VIII, XII, XIII. A still earlier, but more roundabout, transcontinental connection was made in 1881 when a branch of the Santa Fe reached the Southern Pacific line at Deming in southwestern New Mexico.

55. "Majority Report of the Committee on the Pacific Railroad," 40th Congress, 3d Session, *House Reports No. 219,* pp. 22, 1–3.

56. Woodward, *Reunion and Reaction,* Chap. IV.

57. Donaldson, *The Public Domain,* pp. 269–73. Haney, *A Congressional History of Railways . . . from 1850 to 1887,* pp. 19–22.

58. Haney, *A Congressional History of Railways . . . from 1850 to 1887,* Chaps. VII and VIII; Trottman, *History of the Union Pacific,* Chap. X; Daggett, *Chapters on the History of the Southern Pacific,* Chaps. XX, XXI.

59. The figure of 155,000,000 and also the higher estimates are taken from Donaldson, *The Public Domain,* pp. 268–73. The figures previously used for individual grants are consistent with the former figure. The figures for land actually received are taken from Robbins, *Our Landed Heritage,* pp. 222–24, based on a 1923 Report of the Land Commissioner. On forfeiture the principal reference is Ellis, "The Forfeiture of Railroad Grants, 1867–1894," pp. 27ff. For the circumstances of the New Orleans Pacific grant, see Traxler, The Land Grants for the Thirty-Second Parallel Railroad from the Mississippi to the Pacific (MS), Chap. IX. This author gives the land received (p. 223) as about 700,000 acres.

60. Charles Francis Adams, in *Report . . . of the Pacific Railway Commission,* p. 986.

61. The Pacific Railway Commission (*ibid.,* pp. 146–47), attempted to demonstrate this by a table contrasting "Management Contemplated by Congress and the Method Actually Pursued." The Commission's model, however, is based on the unrealistic assumption of bona fide subscription of the entire $100,000,000 of capital stock. Construction mainly with borrowed funds required a top-heavy capital structure. See Donely, The Construction of the Union Pacific Railroad by the Credit Mobilier of America (MS).

If the earnings of the Union Pacific Railroad for the decade 1870–79 are calculated on the basis of the actual cost of construction rather than on the inflated book value, it appears that the average annual rate of return was 11.6 percent. Fogel, The Union Pacific Railroad (MS), Table VIII. The same author (in Table IX) attempts a calculation of the "social return" on the actual investment in construction. By adding to the 11.6 percent an estimate of the effect of the railroad in increasing the value of land along the route, he obtains a considerably larger figure for the return on the investment to the economy as a whole. Even this, as he points out, does not take into account the savings of shippers at the ends of the route or of the government on its own transportation or any political or non-economic advantages.

62. "The act itself was an experiment. It must be considered in the nature of a proposal to enterprising men to engage in the work, for with the extraordinary obstacles in the way there was no security that capital could be enlisted. If enlisted at all it would be only on conditions which would insure, in case of success, remuneration in proportion to the risk incurred." U.S. Supreme Court, 1 Otto 72 (1875).

63. *Report . . . of the Pacific Railway Commission,* p. 982.

64. Galton, "Report to . . . the Committee . . . for Trade and Foreign Plantations on the Railways of the United States."

65. The first point may serve to explain the somewhat surprising indifference to land grants on the part of certain railroad leaders which is described in Cochran, "Land Grants and Railroad Entrepreneurship." The second point was put as follows with respect to the land grant to the Central Pacific: "In the main its effect was to give a grossly excessive and unnecessary profit to a few persons who held most of the stock of the company without having invested any considerable capital of their own." Daggett, *Chapters on the History of the Southern Pacific,* pp. 58–59. It should be added that the obligation to carry government traffic at reduced rates did place something of a burden on the land-grant railroads after the developmental period was over. See also Haney, *A Congressional History of Railways . . . 1850–1887,* Chap. IV.

66. Gates, *Fifty Million Acres,* p. 253. The pioneer study of railroad policy in land disposal was the same author's book on the Illinois Central.

67. A related grievance of the settlers, though not precisely a barrier to settlement, was the practice of a number of the companies of refusing to patent their "earned" lands until they had completed their sale to third parties, thus avoiding payment of land taxation to the local authorities. In a number of cases, settlers were unable to prove their claims to land within the limits of the grants. For the conflict between settlers and railroads, see Gates, *Fifty Million Acres,* Chap. VIII. and Robbins, *Our Landed Heritage,* Chap. XVI.

68. Trottman, *History of the Union Pacific,* pp. 53–54.

69. C. P. Huntington to Senator George F. Edmunds, April 3, 1876, quoted in 44th Congress, 1st Session, *House Report No. 440,* pp. 103–4.

6. THE STATE PROGRAMS OF THE RECONSTRUCTION SOUTH

1. BPW XLIII (1860–61 and 1861–62), p. 7.

2. Black, *The Railroads of the Confederacy,* Chap. 12, "Wartime Construction Programs," and map on end paper.

3. Goodrich, "The Revulsion against Internal Improvements," esp. Table I. Texas had prohibited stock subscription in 1845, and Virginia, in which stock subscription was the principal form of aid, had prohibited loans to corporations in 1850.

4. Goodrich, "Public Aid . . . in the Reconstruction South," esp. Table I. This table gives a conspectus of legislation governing railroad aid from 1866 through 1877, distinguishing for each state the periods *before, during* and *after* Radical Reconstruction.

5. *Report of the Committee to Investigate the Bonds . . . Issued or Negotiated since July 4, 1868,* pp. 5, 183. Italics are in the original.

An early governor endorsed $400,000 of the bonds of the Macon and Brunswick, and endorsements of railroad bonds to the amount of at least $600,000 were made after 1870 on claims based on the legislation of that year. Senate *Journal* (1872), pp. 21–22; (1873), pp. 15–16.

6. *The Evidence taken by the Joint Committee of the Legislature . . . to Investigate the Management of the State Road.*

7. Resolution of August 26, 1872.

8. Tennessee General Assembly, *Appendix to the Senate and House Journals,* 1870–71, pp. 1–255, 327–670, 721–994.

9. "The State Debt Report of the Committee to Investigate It," *Appendix to the House Journal,* 1879, pp. 15–17, 29–30.

10. Nashville, *Tennessee Tribune,* November 13, 1871.

11. Tuscaloosa *Observer,* quoted in Mobile *Daily Advertiser,* August 28, 1867.

12. Montgomery *Alabama State Journal,* April 6, 1873; Mobile *Daily Register,* April 5, 1873. A valuable modern article describes the transaction as effecting "a large saving to the state." Moore, "Railroad Building in Alabama During the Reconstruction Period," pp. 421–41, esp. p. 439.

13. Little Rock, *Arkansas Daily Gazette,* September 15, 1870.

14. Railroad Cos. v. Schutte, 103 U.S. 120.

15. Daniels, *Prince of Carpetbaggers,* esp. Chap. XIII.

16. *Ibid.,* p. 212.

17. General Assembly, *Executive and Legislative Documents,* 1868–69, "Message of Governor W. W. Holden," p. 9.

18. *Ibid.,* Document No. 33, "Report of the Senate Investigation Committee," pp. 1–9, 126–27.

19. *Conditions and Prospects of the Blue Ridge Railroad Company in a Letter Addressed to Governor James L. Orr by J. W. Harrison, President of the Company* (Charleston, 1868).

20. General Assembly, *Reports and Resolutions,* 1869–70, pp. 1591–92; 1870–71, pp. 814–24; 1871–72, "Report of the Financial Investigating Committee." The Charleston and Savannah also received aid in

1869 and secured a postponement of the state's lien. Similar postponement had been voted in 1865.

21. Senate *Journal,* 1871, p. 27.

House, *Debates,* 1870, pp. 68–69.

22. Liberty *Herald,* quoted in Jackson *Mississippi Daily Pilot,* June 8, 1871.

23. Woodward, *Reunion and Reaction,* p. 41. For the C. & O., see Turner, *Chessie's Road,* Chap. IV.

24. Ramsdell, *Reconstruction in Texas,* pp. 307–8. For the vote of the preceding year, see San Antonio *Express,* August 3 and 6, 1870.

25. Jackson (Miss.) *Weekly Clarion,* November 17, 1875; Mobile *Register,* August 21, 1873; Daniels, *Prince of Carpetbaggers,* pp. 216, 220–21.

26. Fayette (Miss.) *Chronicle,* quoted in Jackson *Daily Mississippi Pilot,* June 3, 1871.

27. *The Evidence taken by the Joint Committee . . . to Investigate the Management of the State Road,* p. iii.

28. Goodrich, "Public Aid . . . in the Reconstruction South," Table II. The eleven ex-Confederate states had 26 percent of the nation's mileage in 1865 and only 19 percent in 1879.

7. "THE PATRONAGE OF LOCAL LEGISLATION"

1. Putnam, *The Illinois and Michigan Canal,* Chap. V.

2. John H. Goff, in Charles Moore, ed., *The Saint Mary's Falls Canal* (Detroit: Semi-centennial Commission, 1907), pp. 89–186. In 1933 the United States completed the toll-free Illinois Waterway which takes the place of the Illinois and Michigan. In 1959 a constitutional amendment adopted in New York opened the way for possible transfer of its canal system to the federal government.

3. Whitford, *History of the Canal System of the State of New York,* I, Chaps. IV, XXIII. The short Cayuga and Seneca Canal and the Black River Canal were also retained, the latter largely to supply water for the Erie. As of the end of 1882, when collection of tolls came to an end, the Canal Auditor reported a cumulative operating profit on the Erie of $92,000,000 as against construction costs of about $49,500,000, not including interest. For the state canals as a whole, the corresponding figures were $93,500,000 and $78,500,000. *Ibid.,* II, 1068. For the

Barge Canal, see Noble E. Whitford, *History of the Barge Canal of New York State* (Albany: J. B. Lyon Co., 1922).

4. Goodrich, "Revulsion Against Internal Improvements." Not all these constitutions explicitly prohibited the three alternatives of state public works, state loans, and state stock subscriptions. Except in the case of New York's public works, none failed to prohibit the type of action (if any) that had been practiced in the particular state.

5. Haupt, "Appeal to the People of Massachusetts, 1865," p. 2.

6. Bird, *The Hoosac Tunnel,* p. 24. An earlier pamphlet by the same author was entitled *The Road to Ruin; or The Decline and Fall of the Hoosac Tunnel.*

7. *Statement of Herman Haupt to the Joint Special Committee on Troy and Greenfield Railroad and Hoosac Tunnel, February, 1864,* p. 34.

8. Kirkland, *Men, Cities and Transportation,* Vol. I, Chap. XII, "The Great Bore." The calculation of loss is on p. 432. The yardstick argument was advanced by Charles Francis Adams.

9. Massachusetts Board of Railroad Commissioners, *Second Annual Report,* January, 1871, Table 1, p. ix and Appendix D.

10. Kirkland, *Men, Cities and Transportation,* Vol. II, Chap. XVII.

11. Massachusetts *House Documents,* 1878, No. 333, p. 6, cited in *ibid.,* Vol. II, p. 50.

12. Goodrich and Segal, "Baltimore's Aid to Railroads," Table I. Total aid authorized, 1864–86, was $7,284,000. Parts of two authorizations were unused. The effective aid as received by the road, after allowing for discounts on bond sales and deductions for sinking funds, is estimated as $6,055,267. Table II compares current expenditures and returns, and Table IV shows profits and losses on capital account. Also see above, pp. 84–86.

13. Baltimore City Council, *First Branch Journal,* March, 1863, p. 315. The favorable majority report is on pp. 287–93.

14. ARJ XLII (1869), 457.

15. John W. Garrett, in Baltimore *Sun,* February 15, 1866.

16. E. A. Ferguson, quoted in Hollander, "The Cincinnati Southern Railway," p. 20. "The Railway is primarily a legal triumph." *Ibid.,* p. 62.

17. *Ibid.,* p. 22, note 1, quoting Cincinnati *Daily Times,* November 28, 1868.

18. *Ibid.,* p. 73, note 1, quoting Cincinnati *Commercial,* September 4, 1881. The lessee company went into receivership in 1893.

19. *Ibid.*, p. 83.

20. Goodrich, "Local Government Planning of Internal Improvements," pp. 430–31. The count is for the years 1866–72, inclusive.

21. Pierce, *Railroads of New York*. Appendix Tables 1 and 2 give municipal and state aid in full.

22. *Ibid.*, pp. 51, 52, 42.

23. ARJ XLVI (1873), 1090. See also Indianapolis *Sentinel*, November 16, 1872.

24. Pierce, *Railroads of New York*, p. 22.

25. *Ibid.*, p. 58. Chap. 6 describes the experience of the municipalities in the sale or retention of their securities.

26. Jacklin, Local Aid to Railroads in Illinois (MS). The quotation, from the Quincy *Whig*, appears on p. 26. Appendix Chart I gives full figures.

27. July 15, 1870. Cited in *ibid.*, p. 32.

28. Million, *State Aid to Railways in Missouri*, Chaps. IV and V, and Appendix II, esp. p. 255. The figures of investments and receipts are given in a note beginning on p. 187.

29. Lopata, *Local Aid to Railroads in Missouri*, Chaps. V and VII, and Appendix. For St. Louis County, see p. 48.

30. *Ibid.*, p. 62, quoting St. Louis *Missouri Democrat*, September 24, 1869, and St. Louis *Missouri Republican*, July 24, 1868, and February 21, 1871.

31. See above, pp. 214–16, 225.

32. Merk, *Economic History of Wisconsin*, Chap. X.

33. Daggett, *Chapters on the History of the Southern Pacific*, Chap. II and pp. 128–29.

34. Kirkland, *Men, Cities and Transportation*, II, Chap. XXLV.

35. *Ibid.*, I, Chap. XIV.

Chase, *Maine Railroads*.

36. Portland *Press*, April 30, July 23, August 27, 1867; July 10, 1872, cited in Kirkland, *Men, Cities and Transportation*, I, 480, 483.

37. Cleveland and Powell, *Railroad Promotion and Capitalization*, p. 205. See also Rasmus S. Saby, "Early Railroad Legislation in Minnesota," in *Collections of the Minnesota Historical Society* (Minneapolis, 1908), XV, 1–188, esp. pp. 59, 67, and William Walts Folwell, *A History of Minnesota* (St. Paul: Minnesota Historical Society, 1924), III, 45.

38. Beard, "Local Aid to Railroads in Iowa," pp. 1–34. The author points out that no reliable estimate of total aid has been made. Under

the law of 1876, stock obtained by municipal subscriptions was to be distributed to individual taxpayers.

39. Boyle, "Financial History of Kansas," Chaps. V and VI and Appendices F and G. For details of subsidy to railroads leading to the Gulf of Mexico, see Clark, *Then Came the Railroads*, pp. 77, 165–66. See also William Frank Zornow, *Kansas; A History of the Jayhawk State* (Norman: University of Oklahoma Press, 1957), Chap. XI.

40. "Report . . . of the United States Pacific Railway Commission," III, 1334, 1489–90, 2091–92, 2098.

41. Cleveland and Powell, *Railroad Promotion and Capitalization*, p. 205.

42. Lamar, *Dakota Territory, 1861–1889*, Chap. IV and pp. 201–2.

George W. Kingsbury, *History of the Dakota Territory* (Chicago: Clarke Publishing Company, 1915), I, 614–28. Kingsbury points out (p. 128) that the rural precincts of Yankton County for the most part voted against the subsidy in 1871. The voters of Clay County refused to approve a subsidy.

43. Article XIII, Section 1. The same article was amended in 1918 to authorize the state to "engage in works of internal improvement . . . for the purpose of developing the resources and improving the economic facilities of South Dakota." Both the substance and the language of the amendment are strongly reminiscent of the eighteen-thirties.

44. Cochran, *Railroad Leaders*, pp. 251, 288, 301.

45. James McClellan Hamilton, *From Wilderness to Statehood, A History of Montana, 1805–1900* (Portland, Oregon: Benford and Mart, 1857), pp. 373–75.

46. F. G. Young, "The Financial History of the State of Oregon," in *Quarterly of the Oregon Historical Society*, X (1909), 269–70.

William D. Fenton, "The Political History of Oregon from 1865 to 1876," in *Quarterly of the Oregon Historical Society*, II (1901), 349–51.

Hedges, *Henry Villard and the Railways of the Northwest*, pp. 31–44.

47. *Ibid.*, pp. 25–26, 156–58. For the account of a county bond proposal that was defeated by a narrow margin in 1871, see Mills, *Railroads Down the Valley*, pp. 13–14.

48. Arrington, *Great Basin Kingdom*, Chap. X.

49. Kneiss, *Bonanza Railroads*, Chaps. III, IV and V, esp. pp. 117–18.

50. Howard A. Hubbard, "A Chapter in Early Arizona Transporta-

tion History: the Arizona Narrow Gauge Railroad Company," in *University of Arizona Bulletin,* V (1934), 42. See also pp. 15, 20.

51. Virginia counties subscribed $900,000 and the city of Richmond $2,000,000. Turner, *Chessie's Road,* pp. 62–67 and p. 256, note 27.

James M. Callahan, *History of West Virginia, Old and New* (Chicago and New York: American Historical Society, 1923), esp. I, 447.

52. Law of February 2, 1866.

William Elsey Conneley and E. M. Coulter, *History of Kentucky* (Chicago and New York: American Historical Society, 1922), II, 730, 929–31, 945, 951.

53. Goodrich, "Local Government Planning of Internal Improvements," pp. 432–34.

54. Boyle, "Financial History of Kansas," pp. 102–3, records local government aid of more than $550,000 in the years 1891–1902. Clark, *Then Came the Railroads,* p. 166, says: "As late as 1915, sixteen local units in three Kansas counties aided the Anthony and Northern to the amount of $567,500." For Louisiana, see the same reference and for Indiana see Goodrich, "Local Government Planning," p. 434.

55. Kirkland, *Men, Cities and Transportation,* I, 486–93; II, 350.

56. September 11 and December 18, 1890, cited in John Goodrich, A Short History of the Bangor and Aroostook Railroad (MS). Chase, *Maine Railroads,* Chap. XIV.

57. Mills, *Railroads Down the Valleys,* "Prineville Priority," pp. 121–36. The law was enacted on February 16, 1915.

58. For Chicago, Burlington & Quincy, ARJ XLIII (1870), 986–87. For Michigan Central, ARJ XLIII (1870), 818, and Cochran, *Railroad Leaders, 1845–1890,* p. 196. See also Indianapolis *Journal,* September 15, 1869; Indianapolis *Sentinel,* August 27, 1869.

59. McKay, ed., *Debates in the Constitutional Convention of 1875,* p. 404.

60. *Daily Jacksonville* (Ill.) *Journal,* March 14, 1868.

61. Iowa City *Republican,* December 13, 1865, cited in Beard, "Local Aid to Railroads in Iowa," p. 28.

62. Crawfordsville (Ind.) *Review,* quoted in Indianapolis *Sentinel,* July 8, 1869.

63. New Albany (Ind.) *Ledger-Standard,* December 12, 1880.

64. Beard, "Local Aid to Railroads in Iowa," p. 29.

65. George Washington Case, Jr. to Frederick Billings, July 21, 1873, cited in Cochran, *Railroad Leaders,* p. 287. Hedges, *Henry Villard and the Railways of the Northwest,* p. 26, argues that the decisive

factor was the prospect of making money by land sales in the creation of the new town of Tacoma.

66. ARJ XL (1867), 921, 1129.

67. Cairo (Ill.) *Bulletin,* quoted in Indianapolis *Sentinel,* September 2, 1869.

68. John W. Brooks to James F. Joy, January 28, 1871. Both were officials of the Burlington system. Cochran, *Railroad Leaders,* p. 279.

69. Brooks to Ames, January 23, 1872, quoted by Cochran, *Railroad Leaders,* p. 280.

70. ARJ XLVI (1873), 639.

8. PUBLIC PROMOTION AND PRIVATE ENTERPRISE

1. Julius Rubin, Imitation by Canal or Innovation by Railroad (MS), is a pioneer effort in this field. It finds striking differences in the decisions made by Philadelphia, Baltimore, and Boston, under similar objective conditions, when confronted by New York's success with the Erie Canal.

2. The most notable of the recent quantitative contributions are those of Heath on railroad aid in the ante-bellum South, Pierce on railroad aid in New York state, Segal on the public works of New York, Pennsylvania, and Ohio and on internal improvements in Maryland, and Jacklin on local aid to railroads in Illinois. Segal emphasizes differences between the times and amounts of aid authorized and the times and amounts of aid received and spent by company or contractor. Iowa and post-bellum Virginia, Tennessee, and North Carolina are among the states for which figures on local aid are particularly needed.

3. Cranmer, "Canal Investment, 1815–1860." See also comments in same discussion by Harvey H. Segal.

4. Heath, Public Cooperation in Railroad Construction (MS), p. 253, and "Public Railroad Construction," pp. 40–42. The figures of 55 percent and 75 percent are from the earlier work; the article says that a final evaluation "seems likely to raise substantially" the estimate of 55 percent. The total of public investment given in the article, $140,-000,000, includes "remission of Federal import duties; and estimated realized values of Federal land grants prior to 1861." Texas is not included in these figures.

5. U.S. Bureau of the Census, *Historical Statistics of the United States,* p. 201, Series K 18–27; p. 204, Series K 60–70. The original

source of the capitalization figures is Poor's *Manual*. Aside from questions as to the accuracy of the statistics collected by Poor, a number of adjustments would need to be made to translate the capitalization figure into a satisfactory estimate of the cost of construction: on the one hand, deductions for "water" in the capitalization and for betterments after original construction; on the other hand, additions for investment in abandoned projects and for capital written down in bankruptcies and reorganizations. On the side of the public investments, adjustments should be made for discounts or premiums in the disposal of government bonds. In 1890, when it first becomes possible to compare capitalization with estimates for the cost of road and equipment, the latter figures are considerably lower than the former.

6. The nominal totals were about $280,000,000 before the war and appear to have been between $325,000,000 and $350,000,000 for 1861–90. The latter figure, however, represents mainly dollars of the greenback period and is probably subject to somewhat greater deduction than the former for discounts on state and local bonds.

Recent research has made obsolete the estimates of railroad aid, based mainly on records of the railroad companies, presented in Federal Coordinator of Transportation, *Public Aids to Transportation,* Vol. I, Part I; Vol. II, Pt. I, esp. pp. 56–70, and summary table, p. 101. This gives a gross total of about $422,500,000 for aid under the following categories: loans made by the federal government; loans, guarantees and stock subscriptions made by state and local governments; and contributions to construction made by state and local governments as well as by individuals and corporations. This figure represents only about 70 percent of the total suggested in the preceding paragraph. A part of the discrepancy is caused by the fact that the Coordinator's report, as it points out, does not cover aid given to railroads that were never completed or were abandoned in early decades. Allowances for the difference between the face value and the realized value of government obligations reduce the Coordinator's total to about $408,500,000. Further hypothetical calculations, based on attempts to estimate the terms on which the companies could have raised loans and subscriptions entirely from private sources, lead to the report's final estimate of effective aid in the categories listed above of $232,000,000.

7. While the former of these figures is subject to deduction for discount on bonds, the latter is subject to deduction for such inflation in the capital accounts of the railroads as was illustrated in the cases of the Union and Central Pacific.

8. Heath, Public Cooperation in Railroad Construction (MS), p. 251.

9. Goodrich, "The Virginia System," p. 370 and Table 1.

10. Pierce, *Railroads of New York,* Chap. VI, "The Sale of Stock," esp. pp. 126–127.

11. McGrane, *Foreign Bondholders and American State Debts.* The ante-bellum defaults of Mississippi, Louisiana, the Territory of Florida, and Kansas were mainly on state ventures in banking. Expenditures on improvements played a minor and indirect role in the first three cases. The post-bellum defaults were those of Alabama, Arkansas, Florida, Louisiana, North Carolina, South Carolina, and Virginia. In the last case, the greater part of the debt had been incurred for ante-bellum, not post-bellum, improvements.

12. Hillhouse, *Municipal Bonds,* Chaps. II, VII. The author estimates the total amount of local government railroad obligations involved in defaults as between $100,000,000 and $150,000,000.

13. Pierce, *Railroads of New York,* Chap. V, esp. p. 104, and Appendix Tables 3, 4, and 6.

14. Jacklin, Local Aid to Railroads in Illinois (MS), Chap. VI, esp. pp. 245–46.

15. Lopata, *Local Aid to Railroads in Missouri,* Chap. VI. "Several counties had experiences that dwarf fiction" (p. 123). County judges evaded orders to levy taxes by holding meetings in the woods or under bridges, by resigning their positions and by serving jail sentences. An angry mob killed several judges who had signed railroad bonds regarded as fraudulent. The last railroad aid case was compromised in 1919.

16. Cooperstown *Glimmerglass,* August 1938, quoted in Pierce, *The Railroads of New York,* p. 57.

17. "No community in the state of New York ever repudiated an obligation because of its inability to pay. While many communities were heavily bonded and their taxes burdensome, none was bankrupt." Pierce, *The Railroads of New York,* p. 84. Comparing her findings with those of Pierce and Lopata, Jacklin concludes that the courts were not as lenient with Illinois communities on procedural grounds as with those of New York or Missouri. Local Aid to Railroads in Illinois (MS), p. 213*n*.

18. *Report of the Debates and Proceedings of the Convention for the Revision of the Constitution of the State of Indiana,* p. 677.

19. *Debates and Proceedings of the Maryland Reform Convention to Revise the State Constitution,* I, 132.

Tennessee Constitutional Convention of 1834, *Journal,* pp. 152–54, quoted in Folmsbee, *Sectionalism and Internal Improvements in Tennessee,* pp. 109–10.

20. Lincoln, *The Constitutional History of New York,* pp. 79–80.

21. For New York, Pennsylvania, and Ohio, see Segal, Canal Cycles (MS); for Illinois, Krenkel, *Illinois Internal Improvements,* Chap. VIII; for Baltimore, Goodrich and Segal, "Baltimore's Aid to Railroads," esp. p. 14. In the last case, the tax "ranged from 1½ to 63 cents per $100 of property."

22. ARJ VIII (1839), 61.

23. Virginia, *Senate Journal,* 1877–78, Document No. 24, "Report of the Senate Committee on Finance Relative to the Public Debt." A more unexpected illustration comes from Alabama, after Radical Reconstruction had been decisively rejected, when the Auditor of a Conservative government felt constrained to report that the state "stood almost in the attitude of a beneficiary" of the Radicals' "dangerous policy" of subsidizing railroad corporations. *Report of the State Auditor of Alabama,* September 30, 1877, p. 6.

24. In the case of the federal land grants, fiscal arguments were based on savings in the cost of the government's own transportation and—until the Homestead Act—on the increase in the value of the lands retained by the government.

25. *Journal of the Proceedings of the Convention on Internal Improvements of Maryland,* Resolution No. 7.

Annals, 19th Congress, 1st Session, House Document No. 151, p. 6. ARJ XVII (1844), 151.

Athens (Ga.) *Southern Banner,* July 6, 1853, quoted in Phillips, *A History of Transportation in the Eastern Cotton Belt,* p. 222.

Reubens, Government Financing of Private Enterprise (MS), cites De Witt Clinton as declaring in 1827 that the primary object of all improvements was "the public accommodation and the promotion of internal commerce" and revenue "only a secondary inducement."

26. ARJ XXIII (1850), 553.

27. ARJ XLVI (1873), 639, quoting Cincinnati Board of Trade. ARJ XLII (1869), 456, quoting President John W. Garrett of the Baltimore and Ohio Railroad.

28. ARJ V (1836), 3. The place was Utica, New York.

29. The quotation is from the "Majority Report of the Committee on the Pacific Railroad," 1869, p. 19. Further sophistication may be applied to the Utica formula by noting that governments may often borrow at lower interest rates than private corporations and by taking account

of the possibility that government policy may apply a different time-discount to the future than that set by the market. The argument regarding "inappropriable" gains may be stated alternatively in terms of the net "external economies" created by the improvement. For suggestive use of this latter concept in comparing industries other than transportation, see Albert O. Hirschman, *The Strategy of Economic Development* (New Haven: Yale University Press, 1958).

30. The most comprehensive attempt to appraise the economic effects of the Erie Canal is in Whitford, *History of the Canal System of the State of New York,* pp. 808–907. For definition of the areas directly affected, see Sadove, Transportation Improvement and the Appalachian Barrier (MS). In teaching American economic history, the present author has for many years taken the opening of the Erie Canal as the dividing point between the periods of the "Frontier without the Factory" and the "Frontier with the Factory."

31. Jenks, "Railroads as a Force in American Development."

32. See above, p. 17. Sadove, Transportation Improvement and the Appalachian Barrier (MS), p. 269, and Heath, Public Cooperation in Railroad Construction (MS), p. 99, have commented on the delay in making effective use of the James River–Kanawha route. Canals like the Genesee might have had a considerable period of usefulness if construction had not suffered long interruption because of the exigencies of state finances.

33. Rubin, Imitation by Canal or Innovation by Railroad (MS), presents a closely-reasoned argument to support the conclusion that even as early as 1826 enough was known of the potentialities of the railroad so that Pennsylvania should have deferred its decision to build a canal system. Baltimore businessmen chose the railroad alternative in 1828.

34. "Though parochial interests lingered on . . . they came to be overshadowed by larger, national considerations. . . . Both investors and promoters, now looking far beyond the local market areas, sought the profits to be obtained from the railroad lines which benefited any part of the continent." Taylor and Neu, *The American Railroad Network,* pp. 6–7.

35. *Ibid.,* pp. 8–20, 30–32. Chapter VII is entitled "Solving the Gauge Differentials." See also Kirkland, *Men, Cities and Transportation,* Vol. I, Chap. VIII, "Portland: Rival or Satellite."

36. Hartz, *Economic Policy and Democratic Thought,* pp. 100–3, 160.

37. Goodrich, "Local Government Planning of Internal Improvements," Table II.

38. Goodrich, "The Virginia System of Mixed Enterprise," pp. 357–60.

39. 1 Otto (1875) 72, at p. 80.

40. 5 Ind. (1854) 38.

41. Goodrich, "The Virginia System of Mixed Enterprise," pp. 360, 378–83.

42. J. D. B. DeBow, *Industrial Resources etc. of the Southern and Western States,* II, 458–59.

Goodrich, "Local Government Planning of Internal Improvements," pp. 428–29.

43. Galton, "Report on the Railways of the United States."

44. ARJ XXVI (1853), 346.

45. The Major of Milwaukee, quoted in ARJ XXX (1857), 379. Edward McPherson, *Sale of the Main Line* (Philadelphia: 1857), p. 46, quoted in Hartz, *Economic Policy and Democratic Thought,* p. 169.

46. *North American Review,* XII (1821), 16–17.

47. George Bliss, *Historical Memoir of the Western Railroad* (Springfield, Mass.: Samuel Bowles, 1863), p. 29. ARJ XLI (1868), 608.

48. This argument, with these and other illustrations, is stated in Goodrich, "Public Spirit and American Improvements." Federal Coordinator of Transportation, *Public Aids to Transportation,* II, 66, includes "collective subscriptions to railroad stocks by citizens of communities" as one of its categories of public aid.

49. Callender, in Lambie and Clemence, *Economic Change in America,* pp. 522–24.

50. "The Commonwealth of Massachusetts went to the rescue of one of its railroads last week with the first state subsidy in U.S. history." *Time,* September 1, 1958. "The impetus given motor and air travel by Federal, State and local government promotional programs" is discussed in Interstate Commerce Commission, "Railroad Passenger Train Deficit," *Docket No. 31954,* May 18, 1959, esp. p. 97.

51. Eckstein, *Water Resource Development;* Julian Margolis, "The Economic Evaluation of Federal Water Resource Development." The quotation is from the Flood Control Act of 1936.

52. Consider for example the Cochabamba–Santa Cruz road recently built by the Bolivian government with a loan from the Export-Import Bank. Its purpose, like that of the Erie Canal, was to open up for set-

tlement an area of good agricultural land previously cut off by lack of transportation facilities. But in this case, unlike that of the Erie Canal but like many others in the twentieth century, construction of the improvement itself was accompanied or followed by a number of complementary measures, which included the construction of a sugar mill with government funds, provision of farm machinery and supervised agricultural credit, and planned experiments in the resettlement of Indian communities.

BIBLIOGRAPHY

A study of public promotion necessarily makes extensive use of public documents—the records of Congress (*Register of Debates, Annals, Congressional Globe*), the session laws of the various states, and the proceedings of state legislatures and of the city councils of cities such as Philadelphia and Baltimore. Since all these are readily identifiable, they are not included in the list below. Newspapers and law cases are omitted for the same reason. The *American Railroad Journal,* which under somewhat varying titles played an important role in the industry from its foundation in 1832, and the annual reports of the Virginia Board of Public Works are cited by initials of title, volume, year, and page, for example, ARJ IX (1839), 289; BPW II (1818), 9. With the exception of a few volumes to which only one or two references are made, and which are of little interest to the student of internal improvements, the list below includes all the other works of which direct use is made in the book.

More detailed information on certain aspects of the subject, and additional references, are given in nine articles prepared by the author of the present volume as part of the same research project. They are listed below. For critical comment on recent work in the field, see Lively, "The American System" and Carter Goodrich, "Recent Contributions to Economic History; the United States, 1789–1860," *Journal of Economic History,* XIX (March, 1959), 25–35. Current interest in internal improvements as a subject for university research is suggested by the number of unpublished doctoral dissertations and master's essays included in the list.

Abbott, Frederick K. The Role of the Civil Engineer in Internal Improvements. Unpublished dissertation, Columbia University, 1952.

Adams, Brooks. "Introduction," in Henry Adams, The Degradation of the Democratic Dogma. New York, Peter Smith, 1949.

Adams, Henry. Life of Albert Gallatin. New York, Peter Smith, 1943.

Adams, Henry Carter. Public Debts. New York, Appleton, 1887.

Adams, John Quincy. Memoirs. Philadelphia, J. B. Lippincott & Co., 1874–77.

Affairs of the Union Pacific Railroad Company. 42d Congress, 3d Session, 1873, House Report No. 78. Cited as Wilson Report.

Albright, George Leslie. Official Explorations for Pacific Railroads, 1853–55. Berkeley, University of California Press, 1921.

Ambler, C. H. Sectionalism in Virginia from 1776 to 1861. Chicago, University of Chicago Press, 1910.

American Philosophical Society. Minutes. Vol. I, 1769–70. MS in Library of American Philosophical Society, Philadelphia.

Arrington, Leonard J. Great Basin Kingdom; an Economic History of the Latter-Day Saints, 1830–1900. Cambridge, Harvard University Press, 1958.

[Baldwin, Loammi]. "Board of Public Works in Virginia," *North American Review,* VIII (1818), 1–26.

Beard, Earl S. "Local Aid to Railroads in Iowa," *Iowa Journal of History,* L (1952), 1–34.

Belcher, Wyatt Winton. The Economic Rivalry between St. Louis and Chicago, 1850–80. New York, Columbia University Press, 1947.

Benton, Elbert Jay. "The Wabash Trade Route in the Development of the Old Northwest," *The Johns Hopkins University Studies in Historical and Political Science,* Series XXI, Nos. 1–2, pp. 1–112.

Bird, F. W. The Hoosac Tunnel; Its Condition and Prospects. Boston, Wright & Potter, 1865.

Bishop, Avard Longley. "The State Works of Pennsylvania," Connecticut Academy of Arts and Sciences, *Transactions,* XIII (1907–08), pp. 147–297.

Black, Robert C., 3d. The Railroads of the Confederacy. Chapel Hill, University of North Carolina Press, 1952.

Black, Robert L. The Little Miami Railroad. Cincinnati [18–?].

Bogart, Ernest Ludlow. Internal Improvements and State Debt in Ohio. New York, Longmans, Green, 1924.

Boyle, James Ernest. "The Financial History of Kansas," *Bulletin of the University of Wisconsin, Economic and Political Series,* V (1908), No. 1, 1–178.

Brayer, Herbert O. William Blackmore, a Case Study in the Economic Development of the West. Vol. II. Denver, Bradford-Robinson, 1949.

Brown, Cecil Kenneth. A State Movement in Railroad Development. Chapel Hill, University of North Carolina Press, 1927.

Burgess, George H., and Miles C. Kennedy. Centennial History of the Pennsylvania Railroad Company. Philadelphia, Pennsylvania Railroad Company, 1949.

Cairns, James Pearson. The Response of New Orleans to the Diversion of Trade from the Mississippi River, 1845–1860. Unpublished master's essay, Columbia University, 1950.

Calhoun, John C. The Works of John C. Calhoun. Vol. V. New York, Appleton, 1857.

Callender, Guy Stevens. "The Early Transportation and Banking Enterprises of the States in Relation to the Growth of Corporations," *Quarterly Journal of Economics*, XVII (1902), 111–62. Reprinted in Lambie and Clemence, Economic Change in America, pp. 522–59.

Carey, Mathew. Brief View of the System of Internal Improvements of the State of Pennsylvania. . . . Philadelphia, 1831.

Chandler, Alfred D., Jr. Henry Varnum Poor. Cambridge, Harvard University Press, 1956.

Chase, Edward E. Maine Railroads; a History of the Development of the Maine Railroad System. Portland, A. J. Huston, 1926.

Clark, Ira G. Then Came the Railroads; the Century from Steam to Diesel in the Southwest, Norman: University of Oklahoma Press, 1958.

Clark, Thomas D. The Beginning of the L. & N. The Development of the Louisville and Nashville Railroad and Its Memphis Branches from 1836 to 1860. Louisville, Standard Printing Co., 1933.

Cleveland, Frederick A., and Fred Wilbur Powell. Railroad Promotion and Capitalization in the United States. New York, Longmans, Green, 1909.

Cleveland, John B. Controversy between John C. Calhoun and Robert Y. Hayne as to the Proper Route of a Railroad from South Carolina to the West. Spartanburg, 1913.

Cochran, Thomas C. Railroad Leaders, 1845–1890; the Business Mind in Action. Cambridge, Harvard University Press, 1953.

——— "Land Grants and Railroad Entrepreneurship," *Journal of Economic History*, X (1950), 53–67.

Collection of All Laws and Resolutions of the General Assembly of Virginia Relating to the Board of Public Works; the Report of the Commissioners Appointed to View Certain Rivers within the Commonwealth in 1812. Richmond, 1819.

Corliss, Carlton J. Main Line of Mid-America; the Story of the Illinois Central. New York, Creative Age Press, 1950.

Cotterill, R. S. "Southern Railroads, 1850–1860," *Mississippi Valley Historical Review*, X (1924), 398–405.

Cranmer, H. Jerome. Canal Investment, 1815–1860 (MS). To be published by the National Bureau of Economic Research, New York, in Trends in the American Economy in the Nineteenth Century, Studies in Income and Wealth, Vol. XXIV.

——— The New Jersey Canals: a Study of the Role of Government in Economic Development. Unpublished dissertation, Columbia University, 1955.

Credit Mobilier Investigation. 42d Congress, 3d Session, 1873, House Report No. 77. Cited as Poland Report.

Daggett, Stuart. Chapters on the History of the Southern Pacific. New York, Ronald, 1922.

Daniels, Jonathan. Prince of Carpetbaggers. Philadelphia, Lippincott, 1958.

Daniels, Wylie J. The Village at the End of the Road; a Chapter in Early Indiana Railroad History. Indiana Historical Society Publications, XIII. Indianapolis, 1941.

Davis, Joseph Stancliffe. Essays in the Earlier History of American Corporations. Vol. II. Cambridge, Harvard University Press, 1917.

Debates and Proceedings of the Maryland Reform Convention to Revise the State Constitution. Vol. I. Annapolis, William M'Neir, 1851.

DeBow, J. D. B. Industrial Resources, etc. of the Southern and Western States. New Orleans, DeBow's Review, 1852–53.

Derrick, Samuel Melanchthon. Centennial History of South Carolina Railroad. Columbia, The State Company, 1930.

Donaldson, Thomas. The Public Domain; its History, with Statistics. Washington, Government Printing Office, 1884.

Donely, George Anthony. The Construction of the Union Pacific Railway by the Credit Mobilier of America. Unpublished master's essay, Columbia University, 1958.

Dorfman, Joseph. The Economic Mind in American Civilization. New York, The Viking Press, Vol. II, 1946; Vol. III, 1949.

Dozier, Howard Douglas. A History of the Atlantic Coast Line Railroad. New York, Houghton Mifflin, 1920.

Dunaway, Wayland Fuller. History of the James River and Kanawha Company. New York, Columbia University Press, 1922.

Durrenberger, Joseph Austin. Turnpikes; a Study of the Toll Road

Movement in the Middle Atlantic States and Maryland. Valdosta, Georgia, Southern Stationery and Printing Co., 1931.

Eckstein, Otto. Water-Resource Development; the Economics of Project Evaluation. Cambridge, Harvard University Press, 1958.

Ellis, David M. "The Forfeiture of Railroad Land Grants, 1867–1894," *Mississippi Valley Historical Review*, XXXIII (1946), 27–60.

Evidence Taken by the Joint Committee of the Legislature . . . to Investigate the Management of the State Road under the Administration of R. B. Bullock. Atlanta, 1872.

Federal Coordinator of Transportation. Public Aids to Transportation. Washington, Government Printing Office, Vol. I, 1940; Vol. II, 1938.

Fogel, Robert W. The Union Pacific Railroad: a Case of Premature Entrepreneurship. Unpublished master's essay, Columbia University, 1959.

Folmsbee, Stanley John. Sectionalism and Internal Improvements in Tennessee. 1796–1845. Philadelphia, 1939.

[Gallatin, Albert]. "Report on Roads and Canals," in American State Papers, Class I, Miscellaneous, I, 724–921. Cited as Gallatin's Report.

Galton, Douglas. Report to the Lords of the Committee of Privy Council for Trade and Foreign Plantations on the Railways of the United States. London, Railways Department, Board of Trade, 1857.

Gates, Paul Wallace. Fifty Million Acres: Conflicts over Kansas Land Policy, 1854–1880. Ithaca, Cornell University Press, 1954.

——— The Illinois Central Railroad and Its Colonization Work. Cambridge, Harvard University Press, 1934.

Glazebrook, G. P. de T. A History of Transportation in Canada. Toronto, Carnegie Endowment for International Peace, 1938.

Goodrich, Carter. "American Development Policy: the Case of Internal Improvements," *Journal of Economic History*, XVI (1956), 449–60.

——— "The Gallatin Plan after One Hundred and Fifty Years," *Proceedings of the American Philosophical Society*, CII (1958), 436–41.

——— "Local Government Planning of Internal Improvements," *Political Science Quarterly*, LXVI (1951), 411–45.

——— "National Planning of Internal Improvements," *Political Science Quarterly*, LXIII (1948), 16–44.

——— "Public Aid to Railroads in the Reconstruction South," *Political Science Quarterly*, LXXI (1956), 407–42.

——— "Public Spirit and American Improvements," *Proceedings of the American Philosophical Society*, XCII (1948), 305–9.

Goodrich, Carter. "The Revulsion against Internal Improvements," *Journal of Economic History,* X (1950), 145–69.

——— "The Virginia System of Mixed Enterprise: a Study of State Planning of Internal Improvements," *Political Science Quarterly,* LXIV (1949), 355–87.

——— and Harvey H. Segal. "Baltimore's Aid to Railroads: a Study in the Municipal Planning of Internal Improvements," *Journal of Economic History,* XIII (1953), 2–35.

Goodrich, John B. A Short History of the Bangor and Aroostook Railroad. Unpublished bachelor's essay, Bowdoin College, 1956.

Handlin, Oscar, and Mary Flug Handlin. Commonwealth; a Study of the Role of Government in the American Economy: Massachusetts, 1774–1861. New York, New York University Press, 1947.

Haney, Lewis H. A Congressional History of Railways in the United States to 1850. Bulletin of the University of Wisconsin, No. 211. Madison, 1908.

——— A Congressional History of Railways in the United States from 1850 to 1887. Bulletin of the University of Wisconsin, No. 342. Madison, 1910.

Hanna, Hugh Sisson. "A Financial History of Maryland, 1789–1848," *The Johns Hopkins University Studies in Historical and Political Science,* XXV (1907), Nos. 8–10, 349–482.

Harrison, Joseph Hobson. The Internal Improvement Issue in the Politics of the Union, 1783–1825. Unpublished dissertation, University of Virginia, 1954.

Hartz, Louis. Economic Policy and Democratic Thought: Pennsylvania, 1776–1860. Cambridge, Harvard University Press, 1948.

Haupt, Herman. "Appeal to the People of Massachusetts, 1865," in *Hoosac Tunnel Papers.* Boston, Wright and Potter, 1866.

Heath, Milton S. Constructive Liberalism; the Role of the State in Economic Development in Georgia to 1860. Cambridge, Harvard University Press, 1954.

——— Public Cooperation in Railroad Construction in the Southern United States to 1861. Unpublished dissertation, Harvard University, 1937.

——— "Public Railroad Construction and the Development of Private Enterprise in the South Before 1861," *Journal of Economic History,* X (1950), 40–53.

Hedges, James Blaine. Henry Villard and the Railways of the Northwest. New Haven, Yale University Press, 1930.

Hidy, Ralph W. The House of Baring in American Trade and Finance. Cambridge, Harvard University Press, 1949.

Hill, Forest G. "Government Engineering Aid to Railroads Before the Civil War," *Journal of Economic History,* XI (1951), 235–46.

——— Roads, Rails and Waterways; the Army Engineers and Early Transportation. Norman, University of Oklahoma Press, 1957.

Hillhouse, A. M. Municipal Bonds. A Century of Experience. New York, Prentice-Hall, 1936.

Hollander, Jacob H. "The Cincinnati Southern Railway; a Study in Municipal Activity," *The Johns Hopkins University Studies in Historical and Political Science,* XII (1894), Nos. I–II, 1–96.

Hoyt, William Henry, ed. The Papers of Archibald D. Murphey. Raleigh, North Carolina State Historical Commission, 1914.

Hungerford, Edward. The Story of the Baltimore and Ohio Railroad. 2 vols. New York, Putnam, 1928.

Jacklin, Kathleen Bessie. Local Aid to Railroads in Illinois, 1848–1870. Unpublished master's essay, Cornell University, 1958.

Jackson, W. Turrentine. Wagon Roads West; a Study of Federal Road Surveys and Construction in the Trans-Mississippi West, 1846–1869. Berkeley, University of California Press, 1952.

Jagtiani, H. M. The Role of the State in the Provision of Railways. London, King, 1924.

Jenks, Leland Hamilton. The Migration of British Capital to 1875. New York, Knopf, 1927.

——— "Railroads as a Force in American Development," *Journal of Economic History,* IV (1944), 1–20. Reprinted in Lambie and Clemence, Economic Change in America, pp. 52–68.

Jervey, Theodore G. Robert Y. Hayne and His Times. New York, Macmillan, 1909.

Jordan, Philip D. The National Road. Indianapolis, Bobbs-Merrill, 1948.

Journal of the Proceedings of the Convention on Internal Improvements of Maryland. Baltimore, 1836.

Kirkland, Edward C. Men, Cities and Transportation; a Study in New England History, 1820–1900. 2 vols. Cambridge, Harvard University Press, 1948.

Kistler, Thelma M. The Rise of Railroads in the Connecticut River Valley. Northampton, Mass., Smith College Studies in History, 1938.

Kneiss, Gilbert H. Bonanza Railroads. Stanford University, Calif., Stanford University Press, 1947.

Kohn, David, ed. Internal Improvements in South Carolina, 1817–1838. Washington, 1938.

Krenkel, John Henry. Illinois Internal Improvements, 1818–1848. Cedar Rapids, Iowa, The Torch Press, 1958.

Lamar, Howard Roberts. Dakota Territory, 1861–1889; a Study of Frontier Politics. New Haven, Yale University Press, 1956.

Lambie, Joseph T., and Richard V. Clemence. Economic Change in America; Readings in the Economic History of the United States. Harrisburg, Pa., Stackpole Company, 1954.

Larson, Laurance Marcellus. "A Financial and Administrative History of Milwaukee," *Bulletin of the University of Wisconsin, Economic and Political Science Series,* IV (1908), No. 2, 137–316.

Lincoln, Charles Z. The Constitutional History of New York. Rochester, Lawyers Cooperative Publishing Company, 1906.

Lively, Robert A. "The American System, a Review Article," *Business History Review,* XXIX (March, 1955), 81–96.

Livingood, James Weston. The Philadelphia-Baltimore Trade Rivalry, 1780–1860. Harrisburg, Pennsylvania Historical and Museum Commission, 1947.

Lopata, Edwin L. Local Aid to Railroads in Missouri. New York, Parnassus Press, 1937.

McGill, Caroline E., and others. History of Transportation in the United States before 1860; Ed. by Balthasar Henry Meyer. Washington, Carnegie Institution of Washington, 1917.

McGrane, Reginald C. Foreign Bondholders and American State Debts. New York, Macmillan, 1935.

McKay, S. S., ed. Debates in the Constitutional Convention of 1875. Austin, Texas, 1930.

Majority Report of the Committee on the Pacific Railroad. 40th Congress, 3d Session, Committee Report No. 219.

Marcuse, William. Local Public Finance in Colonial Connecticut. Unpublished dissertation, Columbia University, 1956.

Margolis, Julius. "The Economic Evaluation of Federal Water Resource Development," *American Economic Review,* XLIX (1959), 96–111.

Martin, William Elejius. "Internal Improvements in Alabama," *The Johns Hopkins University Studies in Historical and Political Science,* XX (1902), No. 4, 127–208.

Merk, Frederick. Economic History of Wisconsin During the Civil War Decade. Madison, State Historical Society, 1916.

Meyer, Balthasar Henry. A History of Early Railroad Legislation in Wisconsin. Wisconsin Historical Collections, IV. Madison, 1898.

Miller, Nathan. "Private Enterprise in Inland Navigation: the Mohawk Route Prior to the Erie Canal," in *New York History*, XXXI (1950), 398–413.

——— The Enterprise of a Free People; The Erie Canal and the Erie Canal Fund in the Economy of New York State, 1815–1837. Unpublished dissertation, Columbia University, 1959.

Million, John W. State Aid to Railways in Missouri. Chicago, University of Chicago Press, 1896.

Mills, Randall V. Railroads Down the Valleys; Some Short Lines of the Oregon Country. Palo Alto, Calif., Pacific Books, 1950.

Moore, A. B. "Railroad Building in Alabama During the Reconstruction Period," *Journal of Southern History*, I (1935), 421–41.

Neu, Irene D. "The Building of the Sault Canal: 1852–1855," *Mississippi Valley Historical Review*, XL (1953), 25–46.

Pacific Railway Surveys, *see* Report of Explorations and Surveys . . .

Pearson, Henry Greenleaf. An American Railroad Builder: John Murray Forbes. Boston, Houghton Mifflin, 1911.

Pease, Theodore Calvin. The Frontier State, 1818–1848. Springfield, Illinois Centennial Commission, 1918.

Pennsylvania Railroad Company, Annual Report, V, February 2, 1852; VI, February 7, 1855.

Phelan, Raymond Vincent. "The Financial History of Wisconsin," *Bulletin of the University of Wisconsin*, II (1907), No. 2, 187–475.

Phillips, Ulrich Bonnell. A History of Transportation in the Eastern Cotton Belt to 1860. New York, Columbia University Press, 1908.

Pierce, Harry H. Railroads of New York; a Study of Government Aid, 1826–1875. Cambridge, Harvard University Press, 1953.

Poland Report, *see* Credit Mobilier Investigation.

Poor, Laura Elizabeth. The First International Railway and the Colonization of New England; Life and Writings of John Alfred Poor. New York, Putnam, 1892.

Poor, Henry Varnum. Manual of the Railroads of the United States. New York, H. V. & H. W. Poor. Cited as Poor's Manual.

Poor's Manual, *see* Poor, H. V.

Porcher, F. A. The History of the Santee Canal. Charleston, South Carolina Historical Society, 1903. (Written in 1875.)

Primm, James Neal. Economic Policy in the Development of a Western

State: Missouri, 1820–1860. Cambridge, Harvard University Press, 1954.

Proceedings and Debates of the Constitutional Convention of 1829–1830. Richmond, Ritchie and Cook, 1830.

Putnam, James William. The Illinois and Michigan Canal; a Study in Economic History. Chicago, University of Chicago Press, 1918.

Ramsdell, Charles W. Reconstruction in Texas. New York, Columbia University Press, 1910.

Reed, S. G. A History of the Texas Railroads. Houston, St. Clair Publishing Co., 1941.

Reizenstein, Milton. "The Economic History of the Baltimore and Ohio Railroad, 1827–1853," *The Johns Hopkins University Studies in Historical and Political Science,* XV (1897), Nos. 7–8, 281–370.

Report of the Debates and Proceedings of the Convention for the Revision of the Constitution of the State of Indiana. Indianapolis, 1850.

Report of the Commission and of the Minority Commissioner of the United States Pacific Railway Commission. Bound with 50th Congress, 1st Session, Senate Executive Documents No. 51. Washington, 1887.

Report of the Committee to Investigate the Bonds . . . Issued or Negotiated since July 4, 1868. Atlanta, 1872.

Report of the President of the North Carolina Railroad to the Governor of the State. Salisbury, N.C., 1859.

Reports of Explorations and Surveys to Ascertain the Most Practicable and Economical Route for a Railroad from the Mississippi River to the Pacific Ocean. 32d Congress, 2d Session, II, House Documents No. 91. Cited as Pacific Railway Surveys.

Reubens, Beatrice G. Government Financing of Private Enterprise: New York State, 1776–1846. Dissertation in progress, Columbia University.

Richardson, James D. A Compilation of the Messages and Papers of the Presidents. New York, Bureau of National Literature [1917 ?].

Riegel, Robert Edgar. The Story of the Western Railroads. New York, Macmillan, 1926.

Robbins, Roy M. Our Landed Heritage. The Public Domain, 1776–1936. Princeton, Princeton University Press, 1942.

Rubin, Julius. Imitation by Canal or Innovation by Railroad: a Comparative Study of the Response to the Erie Canal in Philadelphia, Baltimore, and Boston. Unpublished dissertation, Columbia University, 1959.

Russel, Robert R. Improvement of Communication with the Pacific Coast as an Issue in American Politics. Cedar Rapids, Iowa, The Torch Press, 1948.

Sadove, Abraham H. Transport Improvement and the Appalachian Barrier: a Case Study in Economic Innovation. Unpublished dissertation, Harvard University, 1950.

Sanderlin, Walter S. The Great National Project; a History of the Chesapeake and Ohio Canal. Baltimore, The Johns Hopkins Press, 1946.

Segal, Harvey H. Canal Cycles, 1834–1861; Public Construction Experience in New York, Pennsylvania, and Ohio. Unpublished dissertation, Columbia University, 1956.

——— Maryland: a Mixed Enterprise System. Unpublished Manuscript.

Smith, Alfred G., Jr. Economic Readjustment of an Old Cotton State: South Carolina, 1820–1860. Columbia, South Carolina Press, 1958.

[Sparks, Jared]. "Internal Improvements of North Carolina," *North American Review*, XII (1821), 16–38.

Strickland, W. Report on Canals, Railways and Other Subjects. Philadelphia, 1826.

[Sullivan, John L.] "Internal Improvements in South Carolina," *North American Review*, XIII (1821), 149.

Taylor, George Rogers. The Transportation Revolution, 1815–1860. The Economic History of the United States, IV. New York, Rinehart, 1951.

——— and Irene D. Neu. The American Railroad Network, 1861–1890. Cambridge, Harvard University Press, 1956.

Thorner, David. Investment in Empire: British Railway and Steam Shipping Enterprise in India, 1825–40. Philadelphia, University of Pennsylvania Press, 1950.

Traxler, Ralph N., Jr. The Land Grants for the Thirty-Second Parallel Railroad from the Mississippi to the Pacific. Unpublished dissertation, University of Chicago, 1953.

Trescott, Paul B. "The Louisville and Portland Canal Company, 1825–1874," *Mississippi Valley Historical Review*, XLIV (1958), 686–708.

Trotter, Alexander. Observations on the Financial Position and Credit of Such of the States of the North American Union as Have Contracted Public Debts. London, Longman, 1839.

Trottman, Nelson. History of the Union Pacific; a Financial and Economic Survey. New York, Ronald, 1923.

Turner, Charles W. Chessie's Road. Richmond, Garrett & Massie, 1956.
——— The Virginia Railroads, 1828–1860. Unpublished dissertation, University of Minnesota, 1946.
Van Deusen, John G. Economic Bases of Disunion in South Carolina. New York, Columbia University Press, 1928.
Ward, George Washington. The Early Development of the Chesapeake and Ohio Canal Project. Baltimore, The Johns Hopkins Press, 1899.
Weaver, Charles Clinton. "Internal Improvements in North Carolina Previous to 1860," *The Johns Hopkins University Studies in Historical and Political Science,* XXI (1903), Nos. 3–4, 1–95.
Whitford, Noble E. History of the Canal System of the State of New York. 2 vols. Albany, Brandow Printing Company, 1906.
Wilson Report, *see* Affairs of the Union Pacific Railroad Company.
Woodward, C. Vann. Reunion and Reaction. Boston, Little, Brown, 1951.

INDEX

Bei Fragen zur Produktsicherheit wenden Sie sich bitte an:
If you have any questions regarding product safety,
please contact:

Walter de Gruyter GmbH
Genthiner Straße 13
10785 Berlin
productsafety@degruyterbrill.com